ANTHROPOLOGY OF TOURISM

TOURISM SOCIAL SCIENCE SERIES

Series Editor: Jafar Jafari
Department of Hospitality and Tourism, University of Wisconsin–Stout, Menomonie WI 54751, USA.
Tel: (715) 232-2339; Fax: (714) 232-3200; Email: jafari@uwstout.edu
Associate Editor (this volume): James W. Lett
Indian River Community College, USA.

The books in this Tourism Social Science Series (TSSSeries) are intended to systematically and cumulatively contribute to the formation, embodiment, and advancement of knowledge in the field of tourism.

The TSSSeries' multidisciplinary framework and treatment of tourism includes application of theoretical, methodological, and substantive contributions from such fields as anthropology, business administration, ecology, economics, geography, history, hospitality, leisure, planning, political science, psychology, recreation, religion, sociology, transportation, etc., but it significantly favors state-of-the-art presentations, works featuring new directions, and especially the cross-fertilization of perspectives beyond each of these singular fields. While the development and production of this book series is fashioned after the successful model of *Annals of Tourism Research*, the TSSSeries further aspires to assure each theme a comprehensiveness possible only in book-length academic treatment. Each volume in the series is intended to deal with a particular aspect of this increasingly important subject, thus to play a definitive role in the enlarging and strengthening the foundation of knowledge in the field of tourism, and consequently to expand its frontiers into the new research and scholarship horizons ahead.

New and forthcoming TSSSeries titles:

BOROCZ (The State University of New Jersey Rutgers, USA)
Leisure Migration: A Sociological Study of Tourism

PEARCE, MOSCARDO & ROSS (James Cook University of North Queensland, Australia)
Understanding and Managing the Tourism Community Relationship

VUKONIC (University of Zagreb, Croatia)
Tourism and Religion

Related Elsevier Journals

Annals of Tourism Research
Cornell Quarterly
International Journal of Hospitality Management
International Journal of Intercultural Relations
Tourism Management
World Development

ANTHROPOLOGY OF TOURISM

Dennison Nash

University of Connecticut

Pergamon

U.K.   Elsevier Science Ltd, The Boulevard, Langford Lane, Kidlington,
          Oxford OX5 1GB, U.K.

U.S.A.   Elsevier Science Inc., 660 White Plains Road, Tarrytown,
          New York 10591-5153, U.S.A.

JAPAN   Elsevier Science Japan, Tsunashima Building Annex, 3-20-12 Yushima,
          Bunkyo-ku, Tokyo 133, Japan

First edition 1996

Library of Congress Cataloging in Publication Data

A catalog record for this book is available from the Library of Congress.

British Library Cataloguing in Publication Data

A catalogue record for this book is available from the British Library.

ISBN 0 08 042398 1

Typeset by Techset Ltd, Gateshead, Tyne and Wear
Printed and bound in Great Britain by Bookcraft, Bath

TABLE OF CONTENTS

Preface

This book, which explores a subject that has only recently come under anthropological and other social scientific scrutiny, aims to show how anthropologists have approached the increasingly obvious fact of tourism and what they have found out about it. These days, a good many people of the world have something to do with tourism, which makes the potential audience for the book very large indeed. In planning and writing this study I have been acutely conscious of the size and diversity of this potential audience, which includes colleagues in anthropology and the other social sciences and their students, members of what has come to be called the tourism industry and the many others in various cultures who have some sort of stake in tourism. It has been an intriguing task to try to cover the important issues in the field without becoming banal or using too much of the jargon of our trade. I hope I have succeeded.

It shouldn't take long for a reader to sense my enthusiasm for a field of study which I have been privileged to explore in the early stages of its development. Those of us who have been pioneers in the study of tourism have found the air of discovery and sense of freedom associated with being on the frontier enormously exciting. Being on the margin of our professional cultures, with all the looseness of social ties that this implies, we have had an opportunity for adventure which may be even greater than that which is routine among anthropologists who are well-known for their adventurous natures. Perhaps we tourism scholars are naive, but we do not seem to have been affected by the creeping pessimism that has afflicted some anthropologists like Kuper (1994), who think that socio-cultural anthropology, at least, is floundering about without any productive line of development in sight. As far as the study of tourism by anthropologically oriented scholars is concerned, the sky still seems to be the limit; and like the newly converted everywhere, we are always happy to respond to friendly inquiries about our field.

Besides being privileged to participate in an emerging area of study, I have been fortunate to discover the subject of tourism in the most natural of ways, that is, as a simple extension of a career-long interest in strangers such as immigrants, expatriates, refugees and anthropological field workers. The issue of strangers' adaptation, which has always preoccupied me, may not be so acute among most contemporary tourists, but it can serve as one focal point for research on tourism. There are, in addition, a whole range of other issues that resonate with the central concerns of anthropology and the other social sciences. Indeed, I do not believe that I am being overly optimistic when I point out to colleagues who are sceptical about my choice of subject matter that all – or most – of the major questions that have preoccupied sociocultural anthropologists turn up in the study of tourism. Far from being an intellectual dead-end, therefore, the study of tourism, for me, has provided an opportunity to go well beyond my interest in strangers, from which it emerged.

Even though I have had a great deal of freedom to pursue my tourism agenda, it would be a mistake not to mention the professional involvement with others that has shaped my work. First, I would like to express my appreciation to colleagues in the International Academy for the Study of Tourism, a multidisciplinary, multicultural organization of tourism scholars in which cross-fertilization is a fact of life. This organization, the vitality of which is demonstrated during its biannual meetings and in a series of collective publications by members, is on the cutting edge of knowledge in the field. It also is a stimulating haven for tourism scholars whose efforts are not always fully appreciated in their own disciplines.

Secondly, I would like to mention the intellectual and organizational leadership of Jafar Jafari, the past president of the Academy and Editor-in-Chief of the interdisciplinary *Annals of Tourism Research*, as well as the Tourism Social Science Series, of which this book is a part. It probably would be fair to say that Jafari has done more than any other person to make the field of tourism study respectable and to move it forward intellectually. A glance over any issue of *Annals* will give one something of an idea of the magnitude of the tasks that have occupied him.

I would like to express special appreciation to my reader, the anthropologist James Lett, himself a tourism scholar. It was he who raised questions about the central purpose of this book. Should it simply convey the state of the art, or should it also point the way into the future? The second alternative means, of course, that having surveyed the field, the author should descend from an Olympian perch and join the fray. It should not be too difficult for readers to figure out how I've dealt with this crucial question.

This book has profited from the assistance of Michael Gelfand, a graduate student at the University of Connecticut who has criticized the text and introduced me to new realms of computer know-how. It also has benefited from the support of this university where I have spent most of my academic life. Always supportive of my scholarly projects, the University of Connecticut has continued to provide valuable assistance to me and other Emeritus Professors in the belief that we can continue to add something to its academic life. The writing of this book has been made easier in a number of ways by university support.

Finally, I wish to thank the authors of the works that are summarized and criticized at the end of most chapters in the book. A number of these scholars have been kind enough to provide additional information which has helped to round out my treatment of them and their work. I hope that when they read this volume they will be aware of the importance of their contributions, as well as the depth of my appreciation, and that they will agree that this book represents an advance in our understanding of tourism. Such a response would add infinitely to the pleasure I have had in writing this book.

In closing, I would like to invite young scholars in anthropology and other disciplines to take up the study of tourism and pursue that study with the passion and expertise it deserves. It is, after all, this next generation on whom the development of the field depends. Should this book have something to do with attracting such 'new blood', I shall be even more amply repaid for writing it.

Chapter 1

Introduction

Anthropologists have the habit of putting things in context, and in so doing, help to explain those things. In this way, some ritual practice may be seen to be an aspect of a people's religion, their whole way of life, including its history and even the environment in which they live. It is only natural, then, that in an anthropological study of tourism, there should be some concern with the context in which that study emerged and developed.

Serious study of tourism by anthropologists can be traced back to the early 1960s when Nuñez (1963) published an article on week-end tourism in a Mexican village. At that time, tourism was well on its way to becoming one of the major industries in the world. Further, it had begun to play an obvious role in the development of societies in the less developed world in which many anthropologists have carried out their studies. Finally, it was responsible for an enormous amount of contact between societies which was implicated in social change. But despite its significance for developing societies, its importance in the contemporary world, and as it turns out, humankind past and present, the vast expanse of which provides the ultimate framework for anthropological investigations, one cannot help but feel that anthropological research on tourism has not grown at a pace that matches the significance of the subject. There has been substantial growth, yes, but that growth still seems to have been somewhat retarded. In the light of this, an anthropologist cannot help but turn to the social context in which the anthropology of tourism emerged and developed to try to explain why this is so.

We know that large-scale tourism, such as that provided by the institution of the paid vacation (*congé payé*) in France and the Thomas Cook excursions in England, emerged in the Western world in the late nineteenth and early twentieth centuries when the value of work still reigned supreme among the middle classes, the social sector from which most anthropologists have been recruited. For a middle-class person, leisure, unless it had some useful by-product such as restoring a person or main-

1

taining material means of existence, still was thought of as inconsequential or even a little sinful. Anyone studying tourism, which would seem to involve leisure, would therefore be concerning themselves with something that was not to be taken very seriously. This may be one reason for the puzzled or condescending attitude of colleagues that greeted scholars bent on a study of tourism.

It should be mentioned that this kind of reaction was not unique to anthropology. It seems to have greeted early work by other social scientists, although in varying degrees. Geographers appear to have been the earliest to take up the subject, while sociologists and political scientists came on board only a little ahead of anthropologists. But all seem to have encountered some incomprehension about, or resistance to, their work. Referring to political science, for example, Matthews and Richter (1991:122) say, "For years, scholarly research on tourism was seen as 'frivolous' and not appropriate for mature scholars". And Lanfant (1993) paints a bleak picture in recounting the difficulties of bringing the subject of international tourism into the mainstream of French sociology.

A second reason for the lack of enthusiasm for tourism research among anthropologists could have had something to do with the special character of the anthropological quest. An important aspect of anthropological research is the field trip. To go into the field and gather data about a particular (often exotic) people at first hand is almost a sacred feature of anthropological culture. The goal of such a trip is to get to know a people and to convey that knowledge to others. Most students have to negotiate such a field trip successfully in order to get advanced degrees, and famous names like Franz Boas, Bronislaw Malinowski, Edmund Leach, Clifford Geertz and Margaret Mead are identified with the field study of peoples like the Indians of the northwest coast of America, the Trobrianders, the Kachin, the Balinese, or the Samoans. In the field trip, the anthropologist is supposed to live on intimate terms with a particular people – often in difficult circumstances – while getting to know them. This important rite is supposed to involve serious scholarly work. It often does not go easily. Anxiety, rage, accidents, disease and even death are the field worker's lot. The goal of doing serious, scientific work in alien and frequently difficult territory that eventually produces a picture of a people which does full justice to them and to their way of life, is an assumed – even celebrated – aspect of anthropology. This contrasts with a prevalent view of tourists – all of them travelers, some to far-away, exotic locations – who usually are thought to be bent on enjoying themselves, who rarely gain more than a superficial acquaintance with their hosts and who are not very much interested in really understanding them.

It also may be thought by some that the tourists themselves are exploiters or unwitting representatives of exploiting forces such as international hotel chains, airlines or other national or international agencies which have become involved with native populations. To be accused of exploitation is a very black mark indeed for anthropologists. So, in the anthropological community, the study of tourism could be construed as an invitation to guilt by association with things that anthropological work definitely is not supposed to be: the pursuit of pleasure; superficial observation; and the exploitation of peoples. To avoid negative reactions, therefore, it might have been best not to give the impression that one's research had anything to do with something that might be construed as some kind of pleasure junket to far-away places.

To sum up a somewhat speculative overture to this anthropological study of tourism, it has been suggested that certain values in the culture of anthropologists and others, have discouraged tourism study and therefore were responsible for the (in purely scientific terms) unjustifiably retarded rate of growth of such study in anthropology. These values concerned work, the nature of scientific research and the peculiar character of the anthropological endeavor. In order to be taken seriously by other anthropologists, scientists and laymen, it might have been better for scholars to steer clear of the field. For them to have admitted an interest in the study of tourism, as popularly conceived, was to contaminate them and possibly the rest of anthropology. Those of us among the early wave of tourism researchers have periodically been reminded of that.

There is no need here to say whether the views of anthropological colleagues and others concerning the study of tourism are correct. It is enough to know that they are real and, therefore, that they had the potential to affect behavior – in this case, an avoidance of a field of human action that has much to offer social scientists and others interested in the way human beings act. Not all anthropologists nor other social scientists have been turned away from the study of tourism, however. Moreover, it seems evident that the days in the wilderness for students of tourism are coming to a close. All signs now point to the acceptance of tourism study as a part of the anthropological enterprise. Such study, if informed by some new theory on the cutting edge of anthropological or sociological thought, may even have acquired a certain cachet in some circles. Such acceptance, of course, means that anthropologists in tourism studies will eventually have to give up whatever pleasure–pain they derived from being outsiders on the fringe of their professional culture. Then, they will have the opportunity to follow their scientific interests wherever they lead without being preoccupied with the issue of their legitimacy. How their study of tourism progresses will depend not only on these interests, but what they and other social scientists have accomplished so far and on continuing dialogues with subjects, colleagues in anthropology and other disciplines, and those who have practical interests in the field.

Progress in the Anthropological Study of Tourism

The study of tourism emerged not too long ago in the Western world, and the field continues to be dominated by scholars from North America and northwest Europe, whose work, as has been suggested above, was not immediately accepted by colleagues in the social sciences and others outside these fields. An interest in tourism by anthropologists was barely discernable before the early 1970s, but Dann *et al.* (1988) point out that in the period 1974–1986, 45 articles on tourism by anthropologists appeared in the leading journal in the field, the *Annals of Tourism Research*; and Jafari and Aaser (1988) found the number of anthropology Ph.D. dissertations on tourism to be second only to those in economics in a continuously increasing trend in the social sciences through the 1970s and 1980s. Compendia of note, edited by Smith (1977, 1989) and de Kadt (1979), have appeared. Applied research for governments and other clients has taken hold. Full-length monographs, which will be referred to throughout this book, have appeared in anthropology, as well as in the neighboring social sciences. The *Annual Reviews* have given the field a certain legitimacy with articles for sociology (Cohen 1984) and anthropology (Crick 1989). A book about the future of anthropology (Ahmed and Shore; eds. 1995) has included two chapters on tourism. Finally, there is a small, but growing component of tourism studies by anthropologically sensitive scholars from outside of the core areas of North America and northwest Europe, in which such studies began. Tourism studies regularly appear in the Polish journal *Problemy Turystyki* (Problems of Tourism), the Indian *Tourism Recreation Research*, Croatian *Acta Turistica* and some others. There are also occasional monographs (see e.g., Tsartas 1992; Jurdao 1990; Kaur 1985; Singh 1989; Rajotte and Crocombe 1980), a number of which are in English.

Another indication of a developing interest in tourism is the routine inclusion of the subject in national and international conferences such as those of the American Anthropological Association and the International Congress of Anthropological and Ethnological Science. Also, one should note the establishment of anthropologically sensitive institutes such as the World Tourism Organization in Madrid, the *Centre des Hautes Études Touristiques* in Aix-en-Provence, the Tourism Research Institute in Berne and the International Academy for the Study of Tourism in Madrid and London, Ontario.

It is apparent, then, that the anthropological study of tourism, though of comparatively recent date, may now be sufficiently well-established to permit the identification of past and current trends, to begin some critical evaluation, and to offer some speculations about the future. This book is an attempt to share views on these matters not only with colleagues and students, but with a broader audience which has an interest in tourism, an audience which, as has been suggested by the growth of tourism in the world today, ought to be very large indeed.

The Anthropological Approach

Though anthropologists' interests and procedures overlap other disciplines and though they borrow from other scientists in carrying out their work, anthropology has developed a distinctive approach to its subject matter, something of which already has been suggested. It is true that there are a variety of anthropologies, often at odds with each other, but it does seem possible to discern a mainstream which, so far, seems to have united them. It is the minimal essentials of this mainstream that will be considered in this discussion of the anthropological approach, but readers should always keep in mind the considerable diversity within the discipline and the lively disputation among anthropologists about how to approach and depict human subjects (for the uninitiated who would like to explore the field a bit more, see Nash 1993; Borowsky 1994).

If someone were to ask people on the street what anthropology is all about, they probably would get a number of views. One persistent idea is that anthropologists are concerned with ruins like the pyramids of Teotihuacan outside of Mexico City. Another view is that anthropologists are interested in 'primitives' like the so-called Bushmen of the Kalahari desert in southern Africa or the Walpuri of central Australia, who until recently, still roamed about in a hunting and gathering way of life. Yet another popular notion of the field has anthropologists studying fossils and the evolution of ancient humans such as the Neanderthals of Western Europe.

Like the notions of all those blind men touching the elephant, each of these popular views of anthropology contains an element of truth, but it does not give the whole picture, which involves everything human whenever and wherever it occurs. As far as tourism is concerned, it is the field of sociocultural anthropology, which focusses on peoples' behavior, that is most relevant. Anthropologists have used the term culture to refer to the various activities of human groups or societies. Such activities as tourism do not take place in a vacuum, nor are they unrelated, but rather comprise more or less integrated systems laden with the associative and dissociative tendencies that mark all social life. Some anthropologists downplay this integration in light of the considerable movement and change that is taking place in the contemporary world, and the increasingly fuzzy edges of what once may have been more easily distinguishable, better-integrated cultures, but the assumption that a culture is some kind of 'system' of activities of some group, continues to inform anthropological work. Memorable monographs, such as Turnbull's (1962) study of the Pygmies

[margin annotation:] popular misconceptions about anthropology

of the Congo, Lee's (1979) study of the !Kung San of Angola, Namibia and Botswana and Barth's (1961) investigation of the pastoral Basseri of southern Iran, have conveyed an anthropologist's understanding of exotic peoples' social lives. Though anthropology often has been associated with the study of peoples such as these, it does not stop with them. Anthropologists have also begun to investigate people closer to home, such as American college students in New Jersey, who have been studied by Moffatt (1989), the people of the United States, whose culture has been considered by the Spindlers *et al.* (1990), cadets at an American military college whose ritual practices have been investigated by Adams (1993), English villagers who have been studied by Strathern (1981), Rapport (1993) and others, and even the functionaries of big, impersonal Western bureaucracies (see e.g., Herzfeld 1992). There also have been investigations of historic and prehistoric cultures, as in the case of Wallace's (1978) study of Rockdale, an early American industrial village with textile mills powered by water, pre-historic ones such as a *Homo-erectus* camp near Nice, France, that was investigated by the archeologist de Lumley (1969), or the recent past of a village in Burgundy, which has been studied by Zonabend *et al.* (1980).

All of this suggests that the anthropological purview is very broad and long; indeed, it extends to all of humankind. Anything having to do with human beings and their lives is grist for the anthropological mill. That anthropology involves the study of all human beings, not just those in our immediate vicinity or in the contemporary world, is assumed in this anthropological study of tourism.

When pressed, most anthropologists would agree that they are committed to basic canons of science such as objectivity, verifiability, etc., but the objects they study, namely humans, pose certain problems not faced by natural scientists whose procedures are often cited as a model for scientific endeavor. Some scholars (anthropologists among them) think that the problems humans pose are so great as to make it impossible to carry on legitimate scientific work with them. Many think of anthropology as one of the 'soft' sciences that is only a pale facsimile of the hard-nosed natural sciences, and some anthropologists in the so-called interpretivist camp may even think of their work as belonging to the humanities. This is because they believe that the study of humans cannot, and maybe ought not be carried out with the same procedures as the study of, say, particles in physics.

A crucial problem in studying human beings is that they have subjective worlds of conscious and unconscious intentions and meanings which have to be understood if the full picture of human nature and its various manifestations are to be revealed. So, anthropologists in their fieldwork try to understand what the people they are studying are up to, which includes getting at the meaning that is associated with their actions. This task, the accomplishment of which involves overcoming obstacles to what anthropologists have come to refer to as understanding the 'other', is being regarded these days as particularly demanding, and by some, virtually impossible (see e.g., Rosaldo 1984). Traditionally, such an understanding has been gained by participant observation in which anthropologists observe what is going on while taking part in the life of their subjects, and by the use of informants who tell investigators not only what is happening, but also what is on people's minds. It is obvious that if one is too sensitive to the human subject matter one's objectivity as a scientist may be compromised or vice versa, which forces the anthropologist to carry out a delicate balancing act throughout the course of fieldwork and afterward. The point of anthropological work in the field is the formulation, and eventual publication, of as full and accurate a picture as possible of a people and their way of life.

Needless to say, as with all scientific work, a publication is only one of a series of constructions of the human reality which the anthropologist has chosen to study. Unlike some of his colleagues, the author of the book is optimistic about the possibility of grasping and conveying the nature of this reality, a position that has been effectively elaborated by the anthropologist, Spiro (see e.g., Spiro 1992). There is no question, however, that the task often involves significant difficulties which are not always appreciated by other scientists. A special issue of *Current Anthropology* (Fox 1995) offers a particularly illuminating discussion of the issues involved.

An important aspect of the anthropological approach, already mentioned, is to see human beings and their actions in some social or environmental context which has the potential to influence them. Thus, the aboriginal practice of the Shoshone Indians in western North America of maintaining small camps (sometimes composed of no more than a single nuclear family) for most of the year, can be related to a hunting and gathering existence carried out with rudimentary technology in an environment that supported only a few wild animals and edible plants; and American individualism can be viewed as related to a capitalistic mode of production or to advanced industrialization. Anthropologists are also inveterate comparers, with the scope of the comparisons sometimes ranging broadly throughout humankind. Thus, the little bands of aboriginal Shoshone can be compared with bands of other peoples having the same and different modes of subsistence, kinship structures and environments of similar or greater abundance. Making such comparisons poses special problems of developing cross-culturally applicable concepts, theories and methods. If a particular form of kinship, for example cannot be compared with other forms in other cultures, it cannot further an important part of the anthropological agenda. As far as the anthropological study of tourism is concerned, therefore, if a touristic phenomenon cannot be put in context or compared, its study will make only a small contribution to the anthropological enterprise, and one might add, to tourism studies in general.

To summarize, everything human is within the anthropological purview. This includes the subjective side of what for the observer often can be a rather alien human nature, a fact which raises special problems for anthropologists seeking to comprehend those 'others' they happen to be studying. Among sociocultural anthropologists, for whom the study of tourism is particularly relevant, the focus of interest is on peoples' cultures in which socially patterned ways of doing things such as touring are embedded. Such patterns, appropriately localized in specific groups, say, the Japanese, may be compared with the patterns of other peoples, say, of Germans, with the ultimate framework for comparison being all of humanity.

Within this broad perspective, anthropologists have their special preoccupations. As far as the study of tourism is concerned, one of their more important recent concerns has been with culture contact and culture change, particularly where the contact has involved a more powerful, more developed Europe and North America and the less powerful, less developed world of countries such as those in most of Africa or Latin America. As the anthropologist Nuñez (1977:207–210) has pointed out, the study of Western-inspired acculturation or development among the less developed peoples of the world, as, for example in Wolf's *Europe and the People without History* (1982) or Worsley's *The Three Worlds* (1984), has provided fertile ground for anthropological research on tourism. Indeed, it has continued to dominate such study.

It is interesting to look into the emergence of tourism study in anthropology. Often, an anthropologist investigating acculturation or development among some less developed society in the so-called Third World would be struck by the impact of tourism on these people. As a result, early studies of tourism by anthropologically oriented scholars, again as Nuñez (1977:207) has pointed out, often were serendipitous offshoots of other investigations into culture contact and change. They still are. Some anthropologists continue to be surprised to find how important tourism is in the societies they are investigating. If they are not prepared to take this tourism seriously, they will have additional problems in grasping the sociocultural reality they are bent on comprehending.

An Anthropological View of Tourism

In his review article in the *Annual Reviews*, Crick (1989) argues that though the study of tourism is an important field of inquiry, it never can be a unified area of investigation. According to Crick (1989:312), it is not possible to "envisage a change in the fragmentary, multidisciplinary nature of the field". This apparent 'fact of life', which may be as much due to the nature of touristic reality as the variety of approaches to that reality in the social sciences, suggests that, at least in the near term, tourism study will be carried out from many different points of view.

In the *Annals of Tourism Research*, there have been a number of special issues, each of which approaches the subject from the perspective of a particular discipline, each of which, itself, has a variety of points of view. These have included the following: *The Anthropology of Tourism* (Graburn 1983); *The Economics of International Tourism* (Gray 1982); *The Geography of Tourism* (Mitchell 1979); *Political Science and Tourism* (Matthews 1983); *The Social Psychology of Tourism* (Stringer 1984); *The Sociology of Tourism* (Cohen 1979); and especially the encyclopedic overview, *Tourism Social Science* (Graburn and Jafari 1991), in which the approaches of various social sciences having an interest in tourism were represented. There also has been an active give-and-take across disciplinary boundaries (between anthropology and sociology, especially), which anthropology, having perhaps the most comprehensive view of all, can foster. It seems unlikely, however, that whatever interdisciplinary synthesis develops will involve a unified theory of tourism or some specific methodology to study it. For the moment, the best ecumenical projection would seem to be for some overview of touristic phenomena that could lend a measure of coherence to investigations by researchers within and between disciplines.

Perhaps the best way of setting up such an overview is to consider the tourist as the heart of the entire touristic field. It is obvious that tourists come in many forms; and it is true, as Crick (1989:312–314) points out, that there is disagreement over what a tourist is. For example, some scholars (e.g., Dumazedier 1968; Boyer 1972; MacCannell 1976; Urry 1990; Leiper 1992), though differing on specifics, see the tourist as a creature of modern society, which means that the vast array of pre-modern societies would be ruled out for making comparisons and exploring the roots of tourism.

To anthropologists with their broad-ranging interest in human beings, such a cultural limitation on tourists and tourism doesn't feel right. One only has to look at pre-modern societies such as Medieval England, Ancient Rome, Ancient Greece, the Piaroa of South America and even the hunting and gathering San, as Nash (1979a) did, in order to see why. The briefest acquaintance with these societies will show that there are people in them who look very much like the tourists we know. For example, Seneca is quoted by Balsdon (1969:145) as follows concerning obvious touristic practices in ancient Rome:

People set out (from Rome) with no particular objective in view. They wander down the coast. In a purposeless way they go by sea, they go by land, always wishing that they were doing something else. 'Let us go to Campania'. 'No, smart resorts are a bore; rough country is the thing to see. Let us go to Bruttium and see the ranches in Lucania'.

Going further, if we distill out of all these examples – modern and pre-modern – the definition of tourist as a 'leisured traveler', it does, indeed, seem possible (in contrast to the author's earlier argument that tourism was associated with social conditions of higher productivity, [see Nash 1977:36]) to identify tourists and tourism at all stages of sociocultural development.

Travelers at leisure may be more or less specialized in their actions, as, for example mixing business with pleasure. They may be more or less organized, as on an all-inclusive cruise or going it alone. They may be more or less numerous. They may travel far or near, a distinction that has been consistently associated with social status. And they may practice any one of the many alternative forms of tourism that seem to be multiplying as we speak (see e.g., Smith and Eadington, eds. 1992) Yiannakis and Gibson (1992), for example have already identified thirteen different American-based tourist roles, and this is only the beginning. There is considerable variation, of course, within and across cultures, but all of the variants seem to qualify as leisured travelers, and so, according to the view taken here, as tourists.

In *Tourism as an Anthropological Subject*, the author (Nash 1981:462) paid attention to anthropological and scientific strictures in developing his definition of the tourist as a leisured traveler. Since leisure (which may be thought of as freedom from more important cultural obligations such as work) and travel appear to exist in all societies, the anthropologist, using this definition, is free to make touristic comparisons throughout all of humanity, a most desirable prospect for anthropology. Cohen (1974) has made travel an essential element of his definition of tourism. Travel also is universal, but sorting out tourists from other travelers such as pilgrims may be an even tougher job than deciding what constitutes leisure in a particular society. Each of these definitions, the one based on leisure and the other on travel, gets at an essential aspect of the tourist and tourism, but if the furtherance of the work of the social sciences is taken as the main goal, it looks as if more mileage may be gotten out of the leisure-based definition. Moore *et al.* (1995) point out how difficult it is to separate tourism from leisure; and Butler (1989:567), says, "many of the elements found in and influencing leisure and recreation apply equally to tourism". So, from the point of view taken in this book, the investigation of tourism, which is taken to be one kind of activity practiced during leisure time, should pay special attention to its affinity with other leisure practices. This argument will be elaborated further in *chapter 4.*

From this perspective, one can see leisured travelers (tourists), alone or in groups, as the principal actors playing roles in some tourism drama, which can include hosts of various kinds (e.g., hotel employees, shop-keepers, relatives), people who transport and direct tourists (e.g., airline pilots, bus drivers, guides) and those who make it possible for them to tour in the first place (e.g., travel agents, friends and relatives). All of these actors, as well as others touched by them, can become subjects for tourism research. They constitute a veritable cornucopia of touristic 'others', with whom anthropologists are concerned in their studies of touristic phenomena.

The actors of the tourism drama may also be seen to be playing roles in what appears to be a universal 'touristic process', which as Nash (1984:464) points out, can be viewed as "originating with the generation of tourists and tourisms in some home society or sub-society, continuing as the tourists travel to other places where they encounter hosts with a different culture, and ending as the give-and-take of this encounter affects the tourists, those who serve them, and the various societies and sub-societies involved" (one might add that the process continues on with the return of the tourists to their home societies). Furthermore, this process can take the form of a 'touristic system', "which, itself, could be embedded in some broader social context". In our world of multi-national corporations and far-reaching states, that context, which is variable, can be pretty broad indeed.

Considered in this way, tourism looks particularly inviting for anthropological investigation. It appears to exist in all – or most – of the cultures of humankind (in anthropological parlance, it qualifies as a cultural universal or near-universal), which permits those cross-cultural comparisons that are the hallmark of much anthropological work. It is significant in the contemporary world because of the number of people involved and its role in intercultural contact and developmental change. Even with the anthropological understanding suggested earlier, it still is difficult to comprehend why it took so long for this field of study to become established in the social sciences and anthropology in particular.

Dialogues in the Anthropological Study of Tourism

The anthropological study of tourism can be seen as an outcome of dialogues with a number of significant 'others'. Among these others are the subjects of anthropological studies – those who play various roles in the touristic process. To find out about them, researchers might use a variety of approaches. For example, they might use a time–budget study, in which tourists keep track of what transpires throughout their sojourn, as Douglas Pearce (1986) did with tourists in Vanuatu in the South Pacific. They might employ interviews and participant observation to explore the travel habits of young Americans and Canadians in Munich, as did Hartmann (1988). Evans-Pritchard (1989) used informants and participant observation in the delicate task of getting at native American (host) images of tourists. And Pi-Sunyer (1989) used his position as a native of Catalonia in the course of finding out about the native views of tourists and tourism in a small Catalan maritime community.

Studies such as these, give us some idea of the various kinds of interaction with subjects which provide a basis for the anthropologist's constructions of a particular touristic scene. As in all anthropological research, it should be the aim of the researcher to attempt to follow the twin ideals of empathy and objectivity, which may be more easily said than done. The objective should be to find out what tourists, hosts and others in the touristic process are up to in their actions while viewing these people as dispassionately as possible. Needless to say, there is a problem of positioning here, that is, finding the appropriate viewpoint from which a picture of the subjects and their actions should be drawn. In the end, the anthropologist should be satisfied that he or she has given both subjects and science their due.

Interactions between the anthropologist and subjects are only one aspect of the scientific process. There are also dialogues with other scholars which can contribute to the research endeavor. Possibly because they have been exploring a new field of study and on the margin of their disciplines, social scientists interested in tourism have not been too concerned with disciplinary boundaries. They have not been hesitant to mine other colleagues' territory for concepts, theories, or methods. In this, they have been aided and abetted by the leading journal in the field of tourism, *Annals of Tourism Research*, which, as was pointed out earlier, has been relentlessly multi-disciplinary in its orientation.

As far as the anthropological study of tourism is concerned, the boundary between it and sociology has been practically non-existent. This would seem to be due to a natural affinity of the two disciplines in theory and subject matter as well as the fact that some of the leading students of tourism have had broad training in both disciplines. Consequently, the anthropological study of tourism phenomena is informed by both anthropology and sociology, as well as other disciplines. Because of this, as was mentioned earlier, not all of the work mentioned in this book is of strictly anthropological origin. Indeed, scholars who do not consider themselves anthropologists, but who have used something like an anthropological approach to the study of tourism, are frequently mentioned.

At this point in the development of tourism research, anthropologists interested in the subject are reasonably fluent in a variety of disciplines, but unequally so. Their affinity with sociology is clear; so is their lack of sympathy for economics, which, as will be seen later, certainly has made important contributions to the study of tourism. They also seem to have an aversion to the 'business' side of tourism – marketing, management, etc. Work in these fields has rarely been tapped by an anthropologist interested in tourism. One can only lament this fact and hope that these lacunae of anthropological interest and practice will be overcome in the future. Certainly, as far as the study of tourism in the contemporary world is concerned, economic analyses of tourism would appear to be crucial.

Anthropologists and other social scientists, acting as consultants, have had a hand in helping out those who are actively involved in the touristic process, particularly government agencies. So, they have made suggestions about things like the 'carrying capacity' of a particular location, paths to take for 'sustainable' development, ways to increase tourism benefits to host populations, how to increase tourists' and host satisfactions, etc. An example of this kind of work is to be found in de Kadt's edited volume *Tourism: Passport to Development?* (1979), in which a group of anthropologically-oriented scholars offered perspectives on the nature of tourism and on its social and cultural effects in less developed countries. The policy recommendations adopted in the seminar on which the book is based were directed to UNESCO and the World Bank, which had financed the seminar as a means of gaining information for aiding development projects around the world.

Various governments have supported similar research through governmental and para-governmental agencies which make use of applied research. So, Jagusiewicz (1990), of the Polish Institute of Tourism, considered the prospects for development of Poland's Great Mazurian Lakes; and the research department of China's National Tourism Administration (RDCNTA 1989) studied the relationship between tourism and lifestyle changes in contemporary China.

Though dialogues with public and semi-public bodies have played a part in the development of the anthropological study of tourism, there is very little to report where private enterprise is concerned. Travel agents could benefit from anthropological expertise on destination areas or from anthropological insight into the cultural backgrounds of the traveling public, among other things, but so far, there has been very little action in this direction. The field seems to have been left largely to market researchers like Gess (1972), working for a consulting firm, Gaither International, Inc. Her study of Japanese travel habits, particularly to the United States, was commissioned by the U.S. Department of Commerce. Though capable of providing useful information and hypotheses, these studies lack the broader, culturally informed view of the anthropologist and are usually not theoretically oriented.

Why there has been so little dialogue between anthropologists and what might be called the tourist or travel industry is not entirely clear. Are anthropologists, with their aversion to business practices of the establishment, turned off by this kind of thing? Are their patrons and colleagues in the various ivory towers in which most of them work, likely to look down on them for entering the business world? Are business people, who do use marketing research regularly, so short- and narrow-sighted that the kind of picture offered by the anthropologist, which is likely to be broad and long, is of little use to them? Whatever the reason, as will be demonstrated in later chapters of this book, anthropologists would seem to be missing a good bet by not carrying on this kind of applied research.

There are those in the anthropological community who will say that this lack of involvement with business people is all to the good because by getting into the tourism business one opens up the possibility of being 'bought' by one's employer and, hence, the establishment. What do anthropologists do when they make an assessment that a potentially profitable tourism practice almost certainly will end up harming the host-subjects who, according to the *Principles of Professional Responsibility of the American Anthropological Association,* one is bound to side with? Applied anthropologists studying tourism may encounter problems such as this. Often, there are ways out of such a dilemma and other dilemmas associated with their work, but resolving them may not be easy.

There are, indeed, ways to avoid being 'bought' not only by any employer, but by one's subjects, and anyone else having something to do with one's research. Surely, among the most important of these ways for those studying tourism, is to be aware of oneself, one's social background, the particular part of the tourism process in which one is involved and one's place in it. Finally, most intriguing of all, there are ways to use this kind of research to further the anthropological research agenda. All of these matters will be dealt with later in the book.

Next, it should be mentioned that the anthropological understanding of tourism is shaped not only by dialogues with subjects, with the other social sciences and with people with practical interests in the field, but also, of course, with those in the discipline of anthropology itself. As was pointed out earlier, anthropologists have had special preoccupations and points of view, some of which have conditioned their approach to tourism. One point of view has addressed tourism as a fact of acculturation or development, in which people representing different cultures or societies come into contact and change. An example of this is Greenwood's (1977) famous study in which the concept of commoditization was used to describe the tourism-induced transformation of the meaning of the *alarde* festival in the Basque town of Fuenterrabia into spurious superficialities associated with market exchange.

Another anthropological point of view that has shaped the study of tourism, is derived from the investigations of transition rituals by Turner (1969) and others. Here, the destructured experiential state of people involved in such transitions comes to the fore. Though inapplicable to many tourists, this approach to tourist experiences initially excited a good deal of interest. Suitably revised, which would include operationalizing such Turnerian existential concepts as *communitas* and *liminas*, it would seem to have some potential for directing future research.

Finally, though barely off the ground, there is a line of research being carried out from a point of view that is consistent with the conception of tourism as a kind of superstructure. With its origins in a substantial theoretical tradition in anthropology and other social sciences, and associated mostly with materialist (Marxist) analysis, this perspective, if suitably broadened to include various more essential aspects of a society, not only its material base, can open up a line of research leading to the causes of tourism. For example, Cohen (1988b:377–379) has argued that certain social situations in Western societies tend to produce tourists looking for 'authentic' experiences such as buying traditional crafts or associating with 'real' natives. What are the kind of situations in a home society that generate such touristic interests, as compared with, say, those that generate tourists who are content with *ersatz* holidaymaking? The potential of this approach for anthropological research on tourism would seem to merit much greater interest than has been demonstrated so far.

One ought not get carried away, however, with a discussion of theoretical points of view in any discussion of the anthropology of tourism. As Dann *et al.* (1988) have pointed out, with one or two significant exceptions, there isn't a great deal of theory to consider. There still is a lot of plain description (possibly accompanied by moral judgments) going on in anthropological explorations of the subject. This might also be a factor in delaying the acceptance of tourism studies in the discipline. Those anthropologists interested in tourism still are on the fringe of the anthropological community, and the nature of their dialogue with the community reflects this. They still have not completely put aside value judgments (e.g., tourism is bad) for the pursuit of scientific knowledge for its own sake, a position that Jafari (1990) refers to as the "knowledge based platform". Rather, they need to be reminded, as does Greenwood (1989:183) in his reassessment of his earlier analysis of changes in the Basque festival that were influenced by tourism. He says, "Moral anguish was easier to express. But this had negative effects on tourism as an anthropological subject because it did not suggest the ways that tourism offered opportunities for theoretical growth in the core areas of anthropology".

To conclude, it seems appropriate to discuss the transactions among anthropologically sympathetic scholars who have become interested in tourism. Still a comparatively small face-to-face out-group (as far as their disciplines are concerned) on the cutting edge of knowledge, these scholars have developed a camaraderie that could lose some of its emotional thrust as the field grows. For them, the air of discovery is still fresh and the prospects unlimited. The give-and-take with other disciplines is invigorating. Even the pain–pleasure of not being fully accepted in their disciplines and outside can be attractive.

It seems likely that the study of tourism will eventually become a part of the anthropological and other disciplinary mainstreams – with all that it implies for its practitioners. But when that occurs, when specialization and routinization set in and distinctiveness diminishes, this pioneering little band of brothers and sisters may look back with nostalgia to the time when they still were in an academic wilderness trying to figure out the nature of tourism, beginning to develop dialogues with various others in the touristic process and gradually coming to realize the potential of this fascinating field. Perhaps readers of this book will come to appreciate some of the many satisfactions associated with the development of this particular field of knowledge and be moved by the still somewhat pristine enthusiasm among tourism researchers for the subject.

Itinerary

As was mentioned earlier, anthropology, like the other sciences, has its applied and basic sides. The first is responsive primarily to questions concerning practical affairs such as the possible damage by tourism to a people's way of life. When anthropologists are responding to such questions they are tending towards applied research. On the other hand, if the primary orientation of research is with scientific questions such as the way in which a particular form of tourism is generated and shaped by a society, anthropologists may be thought to be pursuing basic research. These two categories of research are not mutually exclusive, but refer to tendencies only.

The itinerary of this book through the field of anthropologically oriented studies of tourism research involves, first of all, attending to the three basic perspectives that have informed the anthropological approach to the subject. These basic orientations have emerged out of long-standing work in anthropology, sociology and other social sciences. They have influenced anthropologists to look at tourism as a form of development or acculturation, as a personal transition and as some kind of social superstructure. Certainly not exhausting the possible ways of looking at tourism, these points of view have both stimulated and constrained the anthropological study of tourism so far. The book will review the work in each of these areas, assess it and suggest its value for future research.

After following the basic scientific part of the itinerary, the applied side of the field will be considered. Anthropologists, as well as other social scientists, already have been engaged in answering practical questions about tourism that they and others want answered. What these questions are, who wants them answered, what the special scientific and ethical questions posed by this kind of research are and how this dialogue fits into the whole anthropological enterprise, are questions that will be dealt with in this latter section of the book.

Throughout, it is hoped that readers will be able to look at some anthropologically oriented studies of tourism in detail. At the end of most chapters there is a summary case study or two which illustrate some of the important points that have been discussed earlier. Here the aim is not only to give a fairly detailed summary of the research, but also to subject it to a critical analysis; and in order to put a human face on the studies, there will be some information about the scholars involved. It goes without saying that the author has found something of value in the work discussed in the summary case studies, and it is hoped that the reader will too.

Tourism has become an important social fact in our world. Anthropologists have been slow to recognize this and to put the study of tourism on the anthropological agenda. There has been a good many growing pains associated with the emergence of this specialized field of investigation, which will be pointed to throughout this book. But the field now seems to be fairly well established and offers exciting prospects for anthropologists and others with an interest in the subject. Having completed our itinerary at the end of the book, perhaps readers will agree with this assessment.

Chapter 2

Tourism as Acculturation or Development

We are told by Harrell-Bond (1978:5) that "tourist agents discovered the Gambia". This little west African country, sandwiched between the northern and southern reaches of Senegal, gained its independence from Britain in 1965. Almost wholly dependent on agriculture (peanuts, mostly) for foreign exchange, the Gambian Government began to consider ways to diversify the economy. The small, but growing tourism sector came to be seen as a serious alternative for Gambian development. Advice for expanding tourism was available from international agencies such as the U.N. and the World Bank, which were intrigued by its potential for improving the lives of people in the Third World and using tourism to do it.

Experts from these organizations, impressed with the possibilities for tourism growth in the Gambia, began to draw up plans for tourism development there. Their recommendations (see Harrell-Bond 1978:6–7) included a number of infrastructural developments such as improved power and sewage facilities, the location of labor support areas near the prime tourist site in a beach area not far from the capital (Banjul), the creation of a hotel training school and the establishment of an international airport capable of handling increased tourist traffic. In 1975, after agreeing to follow these recommendations, the Gambian Government received bank credits for expanding its tourism facilities. The government also had begun to offer various incentives such as tax relief and investment guarantees "for attracting foreign visitors and entrepreneurs to provide tourism facilities" (Harrell-Bond 1978:8).

This is a frequent scenario for the beginnings of tourism development in the Third World, an area of the globe where anthropologists have staked out a proprietary interest through their studies of what some of them have called "the little peoples of the world". Some of these Third World countries were rich in natural resources such as minerals or good agricultural land, which could be used as a foundation for development. Others possessed the 'sun, sea, and sand' that drew Scandinavian tourists to the Gambia. Such resources had to be developed, however, if the tourist or any other trade were to prosper.

A number of big international agencies stood ready to suggest ways to handle that development, contribute the necessary capital to accomplish it and do the actual business of creating and maintaining tourism facilities. Such assistance did not come without strings attached, however. There were, of course, loans to repay, but also the prospect of a growing dependence on the donors and their ways of doing things. In the rush of enthusiasm for possible tourism bonanzas, such consequences often were not seriously considered.

An Anthropological Look at the Problem

How might an anthropologist look at the scenario just described? First, it probably would be seen as involving some sort of sociocultural change. As Bee (1974) has pointed out in his comprehensive treatment of the subject, all sociocultural systems reveal an interplay of change and persistence, but often one of these aspects is more important than the other. Anthropologists have long been interested in the subject of sociocultural change, particularly so, since they became aware of the enormous impact of Western on non-Western societies, as, for example in the imperial thrust of Great Britain around the world. The study of tourism in anthropology began when change-sensitive researchers discovered that tourism was implicated in the developments in some society they had chosen to study. The unplanned-for quality that Nuñez noted in early anthropological work on tourism continues, but now, as Lett (1989:275) notes, anthropologists are "purposefully turning their attention to tourism".

Is tourism development good or bad for the people involved? Jafari (1990) has outlined the positions that have been taken by people concerned with tourism development. One of these positions, the 'advocacy platform', is illustrated in the Harrell-Bond article by the statements of experts favoring Gambian tourism development. These experts, who sometimes recommend what seem to be ethnocentric, Western-oriented strategies, see tourism as ultimately beneficial to the hosts. They think that investors can anticipate a substantial rate of return and that benefits to the hosts will exceed costs. Among those benefits predicted for the Gambian people, according to one expert (Peil 1977), were the wages paid to a growing work force that would rotate between the agricultural sector in the wet season and the tourism sector in the dry season.

Harrell-Bond, on the other hand, is critical of the experts' plans and proceeds to take apart their developmental scenario. She, like many anthropologists, had reservations about development that has been inspired and directed from the outside. According to her (Harrell-Bond 1978:9–10), the calculations of direct and indirect benefits in the experts' plans were spurious. Natural rotation of the work force between tourism and agricultural sectors would not occur. The jobs generated in tourism and beyond would not be all that numerous and would be mostly in lower paying positions. The biggest salaries would be paid to expatriates, which in fact, along with the need to import tourist amenities such as frozen foods and Western-style toilets, would produce a major leakage of income supposedly to be derived from tourism.

Harrell-Bond also points to inevitable social problems, such as begging, stealing and prostitution, which she thinks will be associated with tourism and which already were significant at the time she wrote. Finally, she mentions the dependency relations with outside metropoles which the development plans could promote. Generally, her view, which frequently is encountered in anthropology, conforms to an extreme version of what Jafari has called the "cautionary platform" regarding tourism and other kinds of externally dictated development, which ranges from outright rejection to an abiding mistrust of advocates' claims.

A look at Gambian tourism development with the benefit of hindsight shows that both positions in this Gambian 'debate' were, to some extent, warranted. Tourism has grown dramatically from only a handful in 1965 to over 100,000 arrivals in 1991–1992 when it produced about 10% of the Gross Domestic Product. On the other hand, many of the problems that Harrell-Bond pointed out earlier still were evident. For example, the predicted seasonal rotation of workers has not occurred, leakage of income by the purchase of Western amenities was considerable, benefits still tended to flow toward outsiders and the local elite, control of tourism development often was in the hands of outsiders and petty crime and prostitution (both female and male) were significant (see Diecke 1993:425–433; Farver 1984:252–263).

As far as tourism development is concerned, positions representative of both Jafari's 'advocacy' and 'cautionary platforms' may be justified. It is interesting, therefore, to note that most of those who have looked at the subject from an anthropological point of view have, like Harrell-Bond, adopted the latter position. Loukissas (1978) points to the degradation of the environment on Myconos. Lee (1978) views tourism as enhancing the position of an entrenched elite in Yucatan. Jordan (1980) sees it undermining and distorting traditional values in a vacation village in Vermont. Pi-Sunyer (1977), studying mass tourism on the Costa Brava, thinks that it promotes stereotyping. Kottak (1966) sees a general deterioration of the communal life of a Brazilian fishing village brought about by the influx of second homes, sport fishermen and hippies. Rosenberg (1988) argues that it has contributed to the demise of agriculture in a small mountain village in the French Dauphiné where grazing animals came to be used mainly to 'mow' the newly developed ski slopes. Finally, to close out this by no means exhaustive, but fairly representative list of indictments, there is Greenwood's (1977) initial denunciation of the tourism-induced commoditization of the *alarde* ritual of a Basque community he had been studying. That this lovely old ceremony, so much a part of the community, was becoming an element in modern market mechanisms seems to have been too much for him to bear.

There were, however, a few early researchers, for example McKean (1976) in Bali, Cohen (1979a) in Thailand, Boissevain (1978) in Malta and Hermans (1981) on the Costa Brava, who found tourism to be a benign and possibly beneficial agent of change. Others, for example those attending a seminar convened by the World Bank (de Kadt 1979), thought that tourism could have a range of consequences ranging from good to bad, but with proper intervention, benefits could be made to exceed the costs for host peoples. Applied research, such as impact assessments by Pye and Lin for Asia (1983), Preister (1987) for an American ski resort, and the World Tourism Organization for Bhutan (see Valene Smith 1981) have followed in this vein. Proper intervention, however, has turned out to be a highly problematic thing.

Though early studies in tourism development certainly contributed to a general understanding of the subject, there was a tendency to rush to judgment about the consequences for host peoples, an attitude more appropriate for moral or political tracts such as those of the Ecumenical Coalition on Third world Tourism (see ECTWT 1986; Millman 1988; Srisang, 1989), which thinks that modern, mass tourism (sometimes glossed simply as tourism) is bad for the Third World. A more considered critical position has been taken by a new interest group, Tourism Concern, founded in the United Kingdom, which is part of a global network of organizations concerned with the growth and impact of tourism (see Botterill 1991).

While it may never be possible to eliminate value judgments in scientific work, ideally, they should not overpower dispassionate scientific analysis. As far as tourism or any other subject is concerned, a first step towards legitimate scientific inquiry is to take that subject seriously. Then it should be considered in an objective analytic way. The tendencies to either dismiss tourism or to make snap judgments about its consequences are still with us, but there are a number of indications that more scientific considerations are coming to prevail in anthropological research on tourism, a development towards what Jafari refers to as the "knowledge based platform". A notable example of this is the study by a Korean anthropologist, Moon (1989), who began her investigation of a Japanese mountain village with the notion that external impositions such as the development of skiing would be bad for village life, but in the course of her research, came around to the view that it was, on balance, beneficial.

Greater objectivity in research on tourism development has been helped along by restudies of certain locations or rethinking earlier work. Oglethorpe (1984), for example, sees that, considering the growing 'crisis of dependence' on British tourism, earlier positive assessments of tourism development in Malta may be difficult to sustain; and given a chance to rethink his conclusion on the subject, Greenwood (1989:181–185), sees that the outrage he had felt earlier about the tourism-induced changes in the Basque town of Fuenterrabia, had prevented him from exploring the full extent of those changes.

The nature of the sociocultural and environmental changes induced by tourism depends, of course, on what kind of tourism is involved. Because of sheer numbers, people tend to think first of modern, mass tourism (with all its negative consequences) as a model for all tourism. *The Golden Hordes* by Turner and Ash (1975) is the kind of book that contributes to such a view. There are, however, many different forms of tourism which have various effects on host peoples. An understanding of this has existed in the anthropological literature on the subject for some time. For example, in early papers, Cohen (1979b) and Nash (1981) stressed the variety of tourists, and more recently Crick (1989) has argued in the same vein.

That this awareness of tourism's diversity has become well established not only in anthropology, but also the other social sciences, is demonstrated by the Zakopane seminar of the International Academy for the Study of Tourism (Nash and Butler 1990; Smith and Eadington 1992), which wrestled with the term 'alternative tourism'. The conclusion of the seminar was that there are many different forms of tourism, each of which may be more or less sustainable for host peoples. As a result of their discussions, the participants in this seminar proposed that the term 'alternative forms of tourism' be used in order to get on with the task of analyzing the various kinds of tourism development and their consequences. A subsequent meeting of the World Tourism Organization came to a similar conclusion.

How does tourism induce changes in the host society? Anthropologists have generally assumed that the elements of any sociocultural system are, in some degree, related to each other and that a change in one is likely to lead, sooner or later, to changes in others. Thus, the rule of residence (where the newly married live in relation to their relatives) has been suggested by Murdock (1949) to be the key element of any kinship system. He argues that if it changes, other aspects of kinship will tend to follow until they line up with it. So the development of neolocal residence, in which the newly married tend to live by themselves, will be followed by the disintegration of larger kinship groupings such as clans and extended families, and the nuclear family consisting of only parents and children will come to the fore. The notion of holistic integration of clearly demarcated social groups, at least in the contemporary world, has recently come under attack in the light of the extensive mobility and change that is self-evident, but it continues to pervade anthropological views of human cultures.

With the introduction of tourism or any other new cultural element into a host society, certain forces are set in motion which lead to changes in the host people's way of life. How does this come about? Among the processes suggested is the 'ripple' or 'multiplier effect', a term derived from economics, in which activities in the tourism sector, such as the building of hotels or airports, beget other activities, such as the creation of businesses that sell to workers, etc. The notion of concentric ripple benefits spreading throughout a society often has proved misleading by a disturbing tendency for tourism benefits to spread mostly in the direction of local and foreign-based elites. The 'demonstration effect' is another concept that has been used by anthropologists to refer to a process in which tourists and the things associated with them become models for the hosts. The increasing use of the concept of 'commoditization', in which social relationships are shaped by the dictates of market exchange, given impetus by Greenwood's study of the Basque *alarde* festival, now serves as an important point of articulation with anthropological, sociological and other theories (see Watson and Kopachevsky 1994).

Other concepts in the anthropological lexicon, such as 'internalization' and 'socialization', in which cultural traits become personalized, 'social contradiction' and 'conflict', in which people and cultural elements come to be at odds with one another, 'adaptation', in which people and cultures adjust to the new cultural elements, 'disintegration', revitalization and the like, are available for consideration. Though economists make liberal, often quite sophisticated use of terms such as the multiplier effect in their studies of tourism development, anthropological analyses of the consequences of tourism in destination areas generally are lacking in their application of this or other theoretically derived concepts.

Not only should the social and psychological processes involved in tourism development be understood and fully analyzed, but proving that tourism is indeed, the cause of changes in the host society is required. Here it is necessary to work out a sufficient and necessary link between touristic input, say, the building of a resort complex, with various kinds of sociocultural change, say, the disappearance of larger kinship groupings. To do this, one has to rule out other possible internal and external sources of change such as industrialization, migration and the influence of the mass media, which may be occurring simultaneously and sometimes come in a whole package.

On the whole, anthropologists have been slow to grasp the necessity of this crucial methodological procedure. In the first edition of Smith's *Hosts and Guests* (1977), for example, the authors of the case studies of tourism development and associated sociocultural changes that make up the bulk of the volume confined themselves pretty much to touristic input. In the second edition (Smith 1989), however, there is a gratifying increase among them in the recognition of extra touristic factors as possible sources of change. In her article on tourism development among Alaskan Eskimos, Smith (1989:75–77), herself, now argues that tourism has not been a major agent of change among these people. She thinks, rather, that it has been the development of extractive industries, the contest between the United States and Russia and U.S. governmental policies on welfare that have been mainly responsible. But how does she know this?

Until impressionistic observations, no matter how well based on ethnographic acquaintance, are replaced with methodologically adequate procedures, we will have to take anthropologists' claims about the consequences of tourism development for host peoples as informed hypotheses only. To go beyond such hypotheses, there must be, first, an awareness that some variation of the experimental method of difference is needed. Then, the effects of tourism may be sorted out either by the selection of an appropriate research site, as in cases such as Dominica in the Caribbean, where there is no other input of consequence, or by the manipulation of data, as in statistical analyses of variance.

As has already been intimated, anthropologists have their special ways, which are partly dictated by the nature of their subject matter. So far, methodological rigor does not seem to be one of them. Dann *et al.* (1988:4) do not give the "various ethnographic approaches typically favored by many anthropologists, historians, and political scientists" very high marks in this regard. As noted above, it seems that there has been an increase in sophistication about tourism as a cause of change in host societies, but it remains to be seen whether this sophistication will spread to other methodological operations and how well specifically anthropological concerns, such as giving the points of view of 'others' being studied their due, looking for the relationship of cultural elements to each other, finding the relevant context for the actions being observed and attempting comparisons, can be adapted to more hard-nosed methodological procedures.

Indeed, as was pointed out in *chapter 1*, a current critical wave in anthropology has maintained that the process of understanding others is fraught with so many difficulties that scientific pretensions ought to give way to an 'interpretivist' approach in the humanistic tradition. Certainly, this approach has contributed impressively to our understanding of sociocultural systems, but as the reader of this book should have gathered by now, the author believes that it can only contribute informed hypotheses to anthropological work, which has to be, in the end, scientific.

Guiding Paradigms

On the theoretical–conceptual front, as the title of this chapter indicates, the guiding paradigm for examining tourism's effects on host peoples has been that of acculturation and development. It would be a mistake, however, to take this guidance to involve sophisticated theory testing. One can discern a few more or less explicit theories or fragments of theories, yes, but they are hardly ever adequately tested. Mostly, one gets some statements about social development which are loosely associated with the advent of tourism in a host society.

Though the term has been used more broadly to refer to any socio-cultural change leading to desirable goals, the concept of development in anthropology and the other social sciences usually has been more narrowly conceived in economic terms, as in the growth of the Gross Domestic or National Product or a narrowing of the income gap between the rich and the poor. People tend to think of the developing or less developed world (also referred to as such terms as Third, pre-industrial, or non-Western World) as comprised of countries having low, but often increasing productivity. One way to increase productivity (a goal usually considered desirable) is to obtain capital from the outside. A tourism development project, aided by credit from, for example, the World Bank, would provide such input. Then, if the multiplier effect were to work as projected, the benefits of increasing productivity would spread evenly throughout the society.

But, as in the case of the Gambia, mentioned at the beginning of this chapter, this developmental scenario may be imperfectly realized or subverted. One problem has been the disturbing tendency for the in-putting agencies, which usually act through the local elite, to control the developmental process in their favor. The argument of neo-Marxist dependency theorists, of whom Frank (1972) was an important fore-runner, is that reliance on outside capital leads to dependence, economic leakage, structural inequalities, resentment among the dependent people and skewed economic development has been used as a heuristic or orienting device by some anthropologists (see e.g., Schlechten 1988, discussed in *chapter 8*), but so far it has not been systematically tested on tourism development. Some research by political scientists (see Erisman 1983; Francisco 1983), which provides only a little support for the dependency model in the political arena, could serve as an introduction for anthropologically oriented research on the consequences of tourism in less developed societies.

The concept of acculturation, as used by anthropologists, refers to sociocultural change, desirable or not, that results from culture contact. The Social Science Research Council summer seminar that proposed research on this subject defined acculturation as "culture change that is initiated by the conjunction of two or more cultural systems" (SSRC Seminar 1954:474). One can see the application of the concept in the operation of the demonstration effect, mentioned earlier. Consider a Western tourist from a ship which has stopped for the day at an island in the South Seas. She comes ashore dressed in scanty attire and wanders about the streets of an island town inhabited by people, many of whom have been converted to a fundamentalist brand of Protestantism, and therefore, who are somewhat strait-laced. This tourist and other Westerners like her are one point of culture contact between the islanders and their tourist guests. Now if, as the result of this contact, some women in town were to begin dressing in the Western mode, one might suspect the workings of the demonstration effect and ultimately acculturation.

The SSRC Seminar proposed that research on acculturation consider the nature of the cultural systems involved in the contact, the contact situation, the social relations that follow and the resulting processes of change in the cultures in contact. In practice, anthropologists, including those studying tourism, have tended to focus on acculturation involving more developed Western peoples and less developed native populations, which has mostly assumed some form of dominant–subordinate relationship. Though some promising work has been done on the contact situation, as in Aspelin's (1978) analysis of a Brazilian case of indirect tourism, the various articles on tourism and ethnicity in a special issue of *Annals of Tourism Research*, edited by Keyes and van den Berghe (1984), and MacCannell's (1992) analysis of the "empty meeting grounds" in which tourists from the Western world and their non-Western hosts meet in a post-modern emptiness, anthropologists have mostly been concerned with the consequential changes in host cultures. It is here where acculturation and development paradigms have come together in the study of tourism.

Another area of convergence has to do with the distribution of power in the relationship between the developed and less developed peoples of the world. In places like The Gambia, the tourism developers and their associates had so much power that the hosts could do little more than react to their initiatives. In situations like this, the old saw that beggars can't be choosers is surely applicable. Consider as a prime example Olwig's (1980) account of the natives of St John in the U.S. Virgin Islands, whose traditional way of making a living was effectively eliminated when the U.S. National Park Service, aided by Rockefeller monies, established a national park on about half of the island. These natives earlier had made use of large areas of its territory for grazing and shifting cultivation. With the establishment of the national park, this mode of production was prohibited and many of the natives were forced to look for work elsewhere, including the Rockefeller resort and other tourist sites on and off the island.

Focused as they have been on the less powerful developing world where hosts rarely have much control over tourism development, it has been easy for anthropologists, as Crick (1989:324) has pointed out, to fall into a view of tourism as imperialism and to assume that all tourism involves a dominant–subordinate relationship between cultures. Though it may be useful to argue, as Nash (1977, 1989) has done, that the provision of any service to outsiders is, at bottom, an imperialistic transaction, the assumption that tourists everywhere are able to throw their weight around or that they and their agents from outside can effectively dictate the course of development in some host society is unfortunate for any cross-cultural view of things.

In order to maintain a viable transcultural view of the destination end of the touristic process, it is necessary to remember that most international touristic contact in the contemporary world occurs between cultures in which, as in the case of Europe and North America, the hosts usually are not, on the whole, subordinate. Nor would the traditional dominant–subordinate view of touristic contact apply to those leisure-time visits of members of a band of hunters and gatherers to another band or to present-day people from the country making vacation visits to the city. So, unless the concept is carefully considered and applied, the notion of tourism as a form of imperialism may sometimes be inapplicable in tourism destination areas. It is only that anthropologists' pervading concerns with the underdogs of the world has made it seem so.

So much for the more abstract considerations of this chapter. In the section that follows, it will be possible to see how they are realized in practice by two studies that have been chosen to illustrate some of the points made so far. That the author admires these endeavors goes without saying; but this ought not to deter us from the kind of constructive criticism that will help advance the field of tourism study.

One line of research has been carried out in Andalusia on the southern coast of Spain, the other in the Kingdom of Nepal in the Himalayas. Both have their ear to the ground, so to speak, in regard to sociocultural changes resulting from tourism development. These summary case studies will serve to show not only what research has turned up on tourism development in particular destination areas, but how anthropologically-oriented researchers go about their work on the subject. These comparatively brief synopses and the ones that follow in subsequent chapters certainly are no substitute for the original works, but they should serve to involve readers directly in issues that are raised in this book.

Jurdao Arrones on Residential Tourism Development in Mijas

In a series of provocative works concerning residential tourism development on Spain's Costa del Sol, Francisco Jurdao Arrones has brought the name of Mijas into the center of the debate about the consequences of tourism for local populations. Jurdao is a man of many parts. He has been editor of *Mediterranean* magazine, which discusses tourism in a popularized, but critical way. He also has been comptroller–treasurer of the town of Mijas, which makes him very much an insider as far as local tourism development is concerned. His doctorate is in economics, but he also has lesser degrees in anthropology, administration and journalism. In this part of the world, people are not as specialized as in, say, American academia, but if you pin him down, Jurdao often says that he is an anthropologist. His work certainly reveals much of the anthropological approach that has been discussed earlier.

Mijas, in its broadest extension, is a town or commune that extends about 14 km along the south coast of Spain between Fuengirola and Marbella. Here, a narrow coastal plain lined with splendid beaches gives way rather abruptly to hills and mountains. The village of Mijas (*Mijas casco*), which is the town's administrative hub and main transient tourist attraction, is situated on the steepening slope of the Sierra Blanca and has a wonderful view of the coast and the sea beyond. On a clear day, you can almost see Africa.

The daily tourist round in *Mijas casco* begins with the arrival of tour busses which park near the *ayuntamiento* (town hall). From them, tourists fan out along narrow, often precipitous streets. Donkey or carriage rides are available for those who prefer not to walk. The tourists stop now and then to admire the view, take in some historic site, purchase souvenirs, have something to eat or drink, and, on special days, attend a bull fight in Mijas' tiny bull ring. By the time it gets dark, most of them have gone. Those staying at the local hotel, after enjoying a late meal, may spend some time looking out at the splendid night scene below before going to bed.

It would be interesting to study the impact of all of this transient touristic activity on the village and town, as, for example, in the conflicts over the color of houses, which is not regulated by law, but by a public opinion with touristic appeal (white) on its mind. This, however, is not Jurdao's main concern. He is, rather, interested in the impact of residential tourism, which, for the most part, takes the form of *urbanizaciones* or developments scattered over the landscape usually not too far from the sea. These developments include residences purchased by outsiders who originally came mostly from northern Europe, but now, increasingly, come from other parts of Spain.

Such *urbanizaciones*, which are the work of outside developers who have been assisted in their projects by the financial incentives and a somewhat *laissez faire* attitude of local and national governments, began to appear in the late 1950s. Jurdao (1990), using extensive archival material, interviews and participant observation, gives us an in-depth picture of these *urbanizaciones* and traces the effect they have had on this particular host society in his highly critical work *España en Venta* (*Spain for Sale*).

Before the developers began their work, Mijas was a fairly closed, severely stratified, tradition-bound community where peasants gained a meager existence from small plots of land or by fishing. For the small landholder, subsistence always was in doubt, debt to the rich an inescapable fact and the rules of modern commercial life largely unknown. When pressure began to build for tourism development and buyers came around, peasant owners often made choices they would later regret. Sometimes they sought the advice of the rich and powerful, who, themselves, might be in league with the developers and so were not exactly disinterested. Jurdao cites comments by peasants who got advice to sell their land and have since come to know better.

A typical development would contain a number of homes (chalets or villas) and the necessary infrastructure (streets, lights, water, etc.) to serve them. As in developments in Florida, mentioned by Jurdao, construction often took place at a hectic pace with many deficiencies. The author, who takes a broad, cross-cultural view of such developments, sees the developers as involved in "an enterprise in the process of liquidation", which means that they did the absolute minimum, tried to get away with as much as possible, and sought to get out as quickly as they could.

Because of economic good times in northern Europe, these people had few problems finding buyers. A newly affluent class of workers with guaranteed annual vacations and substantial pensions after retirement could now realize dreams about living in places like the south coast of Spain, which looked very attractive to people who existed in the dark and cold for much of the year. The host people were supposed to be friendly and unspoiled. Crime was non-existent. The cost of living was low, and with improvements in transportation, it was easy to get there and back. Finally, the Spanish Government made it attractive for foreigners to take up residence there. Despite all this, as it turned out, there were problems for buyers, many of them unanticipated.

By 1974, when a good deal of residential construction had been accomplished, the agricultural population of Mijas, which had been approximately 85% in 1950, was reduced to about 7% of the total working population. This change was associated with the displacement from the land of indigenous people, the depletion of the underground water supply (as the result of increased use by new residents of swimming pools, showers, and the like), conflict between residents and locals (over rights of way, noise, etc.), disproportionate benefits to the developers and local elite who took in most of the income from development and increasing control over local affairs by outsiders, as, for example the acquiescence of local and national governments to developers' wishes. Jurdao uses the words "colonization" and "imperialism" to describe what has happened to his township – terms that seem particularly apt in this situation.

Viewed from the host side, which is the perspective favored by anthropologists in such studies, there 'is' a positive side to the development of tourism in Mijas, but, according to Jurdao, it is insignificant. Former agricultural workers do gain employment in construction, and there is a need for maids and gardeners in the developments, but higher paying jobs usually go to better prepared outsiders; and when construction is finished, many workers join the ranks of the unemployed. For these people, going back to one's own land or that of one's relatives usually is not a viable alternative. That land probably has been sold, and the prospects for agricultural work are minimal. The residents in the developments may buy some local goods, but the multiplier effect here is not very strong because they are able to bring in, and buy amenities from home with no or little additional cost, which permits them to maintain old habits of consumption. In short, the spending of foreign residents does not do much to bolster the economy. Neither, apparently, do the fees and taxes, which were absurdly low and were circumvented with impunity by many developers. Further, payment for a residence may have been made directly to a developer in some foreign location and thus did not enter the local economy.

To return to the negative consequences, as viewed by Jurdao, there is the political problem of governing a changing, fragmented society in which neither the local population nor the foreigners are fully integrated. And then there are the not insignificant consequences for the host government in Mijas and other towns along the coast of having an aging, mostly foreign population in their midst. In 1989, more than 50,000 of approximately 67,000 residents of Mijas were foreigners, and of these, approximately half were older than seventy. In *España, Asilo de Europa* (*Spain, Asylum for Europe*), Jurdao and Maria Sanchez Elena (1990) take a look at this population. Using interviews, they find out that these people gripe about, among other things, services (water, telephone, electricity, garbage, mail), noise, the unresponsiveness of the local administration, the flood of tourists into the area and the increased cost of living. Living apart and not knowing the language, they are alienated from their Spanish hosts and incapable of participating knowledgeably in most community affairs. Many of them are lonely, and some drink too much.

In the background of all this, of course, is an awareness by these sojourners that they are getting on in years and that they will have to depend on others to take care of them. But the Spanish health care system is proving to be rather inadequate for handling this aging, increasingly infirm, foreign population. In this regard, of course, Spain is not totally unlike the countries from which these people came. Jurdao and Sanchez (1990) point out that countries sharing the tenets of welfare capitalism are struggling with the facts associated with aging populations. Retirement in them is coming earlier, and it remains for a smaller and smaller working segment of the population to support increasing numbers of people receiving pensions, social security and the like. It is only that Spain has had a major influx of aging, leisured foreigners on their shores and so is struck with this social problem in a particularly acute way.

Though there is a stress on the economic side, the studies of Mijas by Jurdao and Sanchez (1990) are clearly anthropological. First, the people of Mijas and their foreign guests come alive through the investigators' research procedures carried out by natives of the culture being studied. Not all of the people depicted in the studies of tourism development in Mijas seem to be equally real, however. Generally, those in power – the local elite, the developers and the men in Madrid – who initiated and controlled residential tourism development in Mijas were not interviewed and so appear only as vague, sometimes ominous presences, not flesh and blood human beings (Jurdao refers to the 'mafia' and to drug money from time to time, but he does not follow up on these intriguing assertions).

Second, in the Mijas studies, the bigger picture that anthropologists strive for is clearly evident. Tourism development in this community is seen as part of a context, which can be viewed, using the concepts of touristic process or touristic system, as extending to the tourist generating situations in northern Europe and the high councils of Madrid.

Third, cross-cultural comparisons are made with developments in other parts of Spain and elsewhere in which a cyclical, capitalistic scenario, posited by the economist, John Galbraith, appears to have operated. According to this scenario, there is an initial surge of economic activity followed by a decline and then a terminating crash. One is surprised, however, by the lack of reference to some kind of dependency theory in a situation that appears ripe for its use.

Finally, the sticky problem of causation is dealt with here by the use of a site in which tourism seems to have been the only input of consequence. One can be fairly certain that most of the changes in Mijas since the late 1950s have been the result of tourism development. The issue of whether residential tourism or itinerant tourism has been responsible for the observed changes in Mijas is somewhat muted because both forms of tourism development seem to have worked in the same direction. Their cumulative effect, as the author of this book can attest from a personal visit, can cause one to take seriously Jafari's 'cautionary platform'. It is hard to remain scientific about what has gone on in Mijas. Jurdao's critical attitude towards what has transpired in his home town does not seem to be misplaced.

Mountaineering Tourism Development Among the Sherpas of Nepal

It is well known that the ascents of Himalayan peaks such as Mt Everest require the assistance of local guides and porters. The Sherpas of Nepal have become famous in providing assistance in these ascents. In *Tourism and Sherpas, Nepal: Reconstruction of Reciprocity*, Vicanne Adams (1992), a young anthropologist with broad interests that include health, religion and development, investigates the impact of mountaineering tourism on work patterns among the Nepalese Sherpas. A specialist in the Himalayan area, Adams spent 18 months among the Sherpas of Nepal during two periods of field work. She worked with two *cohorts* of over a hundred families each, one living in the Khumbu region near Mt Everest and the other, made up of emigrants from the Khumbu, living in the Nepalese capital, Kathmandu. Information was obtained from a variety of sources. Besides participant observation, Adams and her assistants used semi-structured interviews, follow-up discussions, surveys and governmental statistics, which were employed to the extent that they could be verified and corrected by fieldwork.

Adams tells us that in the days before mountaineering tourism developed in Nepal Sherpa work patterns (in her words, relationships of production) involved a mixture of individualistic wage labor, exemplified by portering, and various hierarchical and egalitarian reciprocities such as *ngalok*, in which goods and services were exchanged over time, and *jindak* bonds between patrons and monasteries. With the introduction of mountaineering tourism under capitalism and the demand for wage laborers, one might have expected an increasing individualism in the relationships of production and a decline in reciprocal labor among the Sherpas. Adams' paper demonstrates that this has so far not been so. Instead, reciprocal relationships have been 'reconstituted' along with the wage labor relationships of mountaineering tourism.

Tourist arrivals in Nepal rose from approximately 16,000 in 1962 to 181,000 in 1985. Trekking agencies have multiplied and jobs in them and related facilities, such as lodging and supply shops, have increased dramatically. Ownership in this 'industry' is in the hands of a few. The major trekking agencies are foreign owned, but Sherpas are well represented in tourism jobs. Indeed, their names are intertwined with most major ascents of Himalayan peaks. Using a variety of examples from the Khumbu village and Kathmandu, Adams shows how new work arrangements have made use of, and extended patterns of reciprocity that formerly were associated with agriculture and religious and secular rituals. For example, she points out that households that cooperated in agriculture and exchanged goods and services were likely to work together in mountain expeditions, that tourism workers were used in the off- (i.e., monsoon) season, possibly as a means of maintaining their availability, that religious and secular rituals in the off-season functioned to strengthen tourism work relationships, that agency owners hired employees on the basis of social connections and that there were exchanges of employees and clients by trekking agency operators.

What may be the most creative use of the old pattern of reciprocity in the tourism business was the institution of a new *jindak* system, formerly existing between monasteries and large donors, in which foreign tourist-climbers were recruited as sponsors. These sponsors, many of who have been caught up in the romance of mountaineering and the personal ties involved in mountain expeditions, might be recruited as benefactors. According to the author, it was normal for a Sherpa to have one large and several minor sponsors, all of them unknown to each other. These sponsors, enthusiastic about their adventures in the Himalayas and grateful to those who had helped make them happen, gave money, gifts, and sometimes, trips to a sponsor's home. In return, they were feted on return visits to Nepal, given fictive kin titles, and generally made to feel a part of a people they had come to admire.

The discussion of foreign sponsorship gives Adams the opportunity to answer the important question concerning how the Sherpas have managed to exert so much control over their tourism rather than being controlled by it, as were the people in Mijas. The persistence of the pattern of reciprocity in their productive relations appears to be part of a broader pattern of cultural persistence, in which, according to Stevens (1993:423), "Sherpas continue to maintain many basic features of their culture and social organization despite their increasing involvement in tourism and other changes in Khumbu life during the past forty years".

One reason for this persistence, which Adams fails to mention, might have been the strength of social reciprocity bonds in the host culture before the advent of tourism. A second reason might be that mountaineering tourism in Nepal still has not attained dimensions which could cause a serious break with the past. Still another reason, as she points out, could be that mountaineering tourism has a nature that 'requires' personal bonding. Rather than being engaged in the more impersonal work of tourism that exists on the Costa del Sol, the Sherpas have personal, reciprocal relationships with each other, and some expedition personnel develop them with their trekking or mountaineering clients. Hence, the possibility of 'reconstructing' reciprocity.

Adams' article probes deeply into the lives of her subjects. One gets the feeling that she really knows the Sherpas and has fully grasped what is happening to them under tourism. It is not that they have been learning capitalistic ways bit by bit, nor that they have been bowled over by powerful capitalistic forces acting through tourism, as seems to have happened in Mijas, but rather that they have managed to absorb the new ways into older patterns of reciprocity. One Nepalese informant who has visited the U.S. sums up the difference between cultures as "the social thing". Instead of simply paying to make things happen, the Nepalese have continued the strategy of creating social obligations.

One wonders, however, how long this can continue. The dominant trekking agencies are, according to Adams, owned by foreigners, and as Stevens (1993:411) points out, foreign ownership in the growing hotel sector is increasing. These surely constitute important entry points for world capitalism, which tends towards more individuated, impersonal operations. It would have been interesting if Adams had managed to compare wage labor and practices in them and possibly other businesses where capitalistic ways are more pronounced with those in which reciprocal ways are more important. Such a comparison would have permitted Adams to translate her rather abstract references to capitalism into real human terms and take a crucial step towards predicting how long the Nepalese strategy of creating social obligations in developing capitalism will be successful.

Finally, it should be pointed out that reciprocity in work patterns is not unique to Sherpas. Adams is aware of this, and at one point she adopts a cross-cultural point of view to suggest that this pattern has been associated with southern Asia, and indeed, that it probably is universal. If so, what she has done in her study is provide one of a variety of instances of more autonomous development in the face of capitalist, and possibly industrial pressures from outside.

* * *

The summaries just given are inevitably inadequate. There is no substitute, of course, for reading the studies themselves. Hopefully, however, they do distill the essentials of each of these important investigations of the impact of different kinds of tourism on host societies in Europe and Asia. In Mijas on the Costa del Sol, the advance of tourism seems to have swept all before it. The hosts there seem to have had little control over the process of tourism development. Among the Sherpas of Nepal, on the other hand, the development of mountaineering tourism appears to have been infused with innovative versions of old host patterns of reciprocity. These hosts seem to have been more active participants in the tourism developmental process and to have put a greater imprint on it.

The researchers involved in these studies, like many others working in less developed societies, have been struck by the dramatic sociocultural changes associated with tourism development and have described them in detail. While doing this, they have continued an anthropological tradition of identifying with the hosts, or at least some of them, a fact which seems to have colored their findings. Adams appears to be enthusiastic about the Sherpas' creative response to powerful outside forces. In spite of all that is happening to them, it seems that they have managed to maintain an essential feature of their culture. Jurdao's response is to react with anger over the way in which outsiders have taken over his native area. Such responses, which take the side of one's subjects, particularly those in the Third World, are typically anthropological.

The two studies just reviewed also are typical of the discipline in not worrying overmuch about whether it is tourism, some specific form of tourism, or something else that is reponsible for the changes observed. Fortunately, it looks as if residential tourism is in fact the major efficient cause of the lamented changes in Mijas, and mountain tourism does appear to have been responsible for contributing to the reconstruction of reciprocity among the Sherpas. Despite this fortunate turn of events, one might have hoped for a little more methodological rigor by the authors in sorting out the issue of causation.

Another fault in both investigations is a failure to extend their analyses adequately along the power trail of the touristic process or system which operated in each situation. In Mijas, it is clear that trail leads into the board rooms of developers and high councils in Madrid where decisions were taken regarding residential tourism development. Jurdao does mention these 'others' who were active in the Mijas developmental scenario, but he never interviews them and so fails to get into their heads, so to speak. They only operate as abstract economic men are supposed to act.

The same problem exists in Adams' work. As mentioned above, it would have been instructive to learn more about those foreign owners of the dominant trekking agencies, and other businesses under greater capitalistic influence and what they were up to, rather than having to entertain abstract notions of capitalistic forces at work. One cannot do everything of course, but it seems essential that investigators identify the principal actors in any tourism activity they choose to investigate and get information directly from them about their view of things, which is to say that they must identify the relevant actors and their interests and thus, the operative field of tourism development.

Next, there is the big question that buzzes in one's mind after having compared these two interesting studies. It is why one host culture gave way and the other did not. Does it have to do only with the magnitude of touristic impact? Certainly, tourism development on the Costa del Sol was a different order of magnitude than in Nepal, and size and rate of development have been cited by the UNESCO and World Bank Seminar (see de Kadt 1979:339–341) as significant factors in the process. Another factor, which undoubtedly overlaps that of magnitude and was discussed by the seminar, is the 'kind' of tourism involved. Mountaineering tourism, although sometimes of considerable scale, is (at least in later stages of the ascent) more likely to be a smaller-scale, collaborative endeavor; in contrast, people living in a chalet or villa on the Costa del Sol are anonymous parties in some large-scale developmental scheme. Finally, there is what may be the most interesting factor of all: the self-determining character of the host–recipient culture. Anthropologists have been understandably interested in those native cultures that manage to resist outside ways, adapt them to their own designs, or use them to revitalize their culture, as nativistic movements, such as the 'ghost dance' of the Sioux Indians are supposed to do. These studies of tourism's impact fit into a line of research that Linton did much to stimulate in his edited volume *Acculturation in Seven American Indian Tribes* (1963).

Where different forms of tourism are concerned, Young's (1977) early cross-cultural study of Caribbean tourism development continues to be instructive. Her statistical analyses suggest, in opposition to any easy notions of imperialism, that the political and economic structures of those societies seem to have influenced the kinds of tourism development that occurred in them. In concluding this important study, Young (1977:672) asks a question that needs to be kept in mind by researchers investigating the impact of tourism on host societies and their environment: "Under what circumstances can a new industry (tourism) that will bring about such fundamental changes be introduced and survive in a host country?"

Another aspect of the tourism development scenario that anthropologists have touched on only lightly is the tourism sector, which will be a more or less autonomous sub-culture according to the degree of social differentiation prevailing in the destination area. That this sector can have its own imperatives, which dictate the course of its own development, is suggested by Butler's (1980) "cycle of evolution", in which a tourist area tends to pass through certain stages, and Oppermann's (1993) "phases" of development of tourist "space". Though possibly giving too much of an impression of inevitability, these schemes do offer considerable leeway for outside input, a fact which is stressed in Nash's (1979b) analysis of the rise and fall of an aristocratic tourist culture in Nice and Brown's (1985) study of the early development of Weston-super-Mare, a British seaside resort. Further studies of this tourism sub-culture, such as that of campgrounds on the South Carolina coast by Janiskee (1990), of the pilgrimage and touristic side of Lourdes by Eade (1992), of low-income plotland and holiday camps in England by Hardy (1990) and of farm tourism in New Zealand by Philip Pearce (1990), are needed.

Undoubtedly, now that anthropologists have discovered tourism and are accepting it as a legitimate subject, there will be an increasing number of investigations at the host or destination end of the touristic process, which, as Dann *et al.* (1988) have pointed out, need to be carried on with increasing methodological sophistication and theoretical awareness. The kind of acculturation situation they have chosen to study involves large power differentials between more and less developed societies mostly in the contemporary, increasingly integrated, capitalist world. That, as has been suggested here, is only one of a variety of acculturation situations associated with the touristic process. Other touristic situations in the socialist world, between developing and developed societies (rural to urban, for example), developed societies (the U.S. and Europe, for example) and less developed societies (contact, say, between bands of hunters and gatherers) in the present and past call for investigation.

This is not all. A focus on the impact of tourism on hosts, their societies and environment leaves out what happens to the tourists themselves and the societies that create them and send them abroad. Furthermore, the processes that generate tourists and tourisms need to be studied, which would help anthropologists and other social scientists to get at the cause, or causes, of tourism rather than only its consequences. Such issues will be considered in *chapters 3* and *4* which follow.

Chapter 3

Tourism as a Personal Transition

Though there may be differences about specifics, everyone seems to agree that tourists, like traders, explorers, or military invaders, are some kind of traveler or visitor. It is this travel that brings about the culture contact discussed in *chapter 2* – the tourists and those associated with them (transport personnel, guides, etc.) being the agents of that contact. Thus, in a widely used definition, the International Union of Travel Organizations (IUOTO 1963:14) referred to the tourist as a certain kind of "temporary visitor". An archetypical Western model for such a traveler–visitor has been the young English gentleman who, beginning in the sixteenth century, made the 'grand tour' of the European continent, during which he added to his formal education and amused himself.

A more egalitarian form of this touring exists today in the form of overseas study programs. People associated with these programs often cite personal development as one of the principal goals of a period of study abroad. Thus, Coelho (1962:66) says that "cross cultural education... is committed to facilitate the process of education of the whole person as an international student and as an individual". Do these programs, in fact, create the experience that leads to such an outcome? Are these young people significantly changed by going abroad, living with their hosts, studying and amusing themselves in a variety of ways? The evidence available so far (see Church 1982) is equivocal and more research is needed. Assuming that students who study abroad have many qualities similar to tourists, information about them can be used in discussing tourists generally.

To carry out investigations of the experience and reactions of tourists, one takes tourists, rather than the hosts or touristic others to be the principal actors in the tourism process. There is no shortage of claims, including scientific ones, about what happens to tourists during their travel–sojourn. For example, on looking through some European and American promotional literature, one encounters statements suggesting that a vacation somewhere can be diverting, relaxing, restoring, broadening, fulfilling, challenging and exciting, among other things. What is the nature of tourist experiences and do they, in fact, lead to such personal consequences? If so, are they significant for an individual's enduring character and for their society? In this chapter, a critical look will be taken at anthropological approaches to such questions and what answers already have been suggested.

Touring and the Tourist

Though travel or migration are ubiquitous among human beings, and though they are instrumental in bringing about culture contact, anthropologists, until recently, have not shown a very great interest in this subject, especially the psychological side of it (see e.g., Harrell-Bond 1992:9–10 concerning the emerging study of refugees). Even the disturbance of culture shock, which is associated with culture contact and which was first popularized by anthropologists (see Oberg 1960), has not received much attention. Sociologists have shown a greater interest. Among the works that have explored the psychological dimension of social change among migrants are those by Thomas and Znaniecki (1927) on Polish immigrants to the United States, and Marris (1975), who sees migrants as one kind of people experiencing loss and change. In that discipline, the concept of the stranger, originating with Simmel (1950) and elaborated on the personal–experiential side by Schuetz (1944, 1945), has gained some currency. In another neighboring discipline, psychology, there is a significant body of research on what Church (1982) refers to as "sojourner adjustment".

For whatever reason, any interest in the nature and consequences of the tourist experience for the tourist could not profit from a well-developed corpus of anthropological knowledge on the subject. After casting about for guiding concepts, Jafari (1987) comes up with the metaphor of the springboard, in which the tourist moves from a state of involvement in ordinary life through emancipation and animation in the realm of "non-ordinary flotation" and back again to the ordinary. Graburn (1989), in a well-known statement that would probably be news to certain Calvinists, for whom the work of the world is central, sees tourists entering a non-ordinary, sacred "high" before coming down (and back) to an ordinary, profane, workaday existence. Such a process is not unlike what is supposed to happen to people in certain transition rituals, including pilgrimages, which have attracted some anthropological interest. What might be referred to as a ritualization of tourism was the result of the application of this religiously-related paradigm to tourism study.

The first extended treatment of transition rituals was by van Gennep (1960[1908]), whose cross-cultural investigations gave some indication of the nature and range of such rites. His work was given more subjective depth by Turner in his extensive studies of ritual processes (see e.g., Victor Turner 1969), which were later to include a pilgrimage (with Edith Turner 1978). In Turner's view, which follows directly from van Gennep, people involved in such rituals go, first, through a process of 'separation' in which they are distanced and freed from ordinary or routine social life. Next, they enter a period of 'liminality' in which the structured necessities of ordinary life dissolve into a destructured, non-ordinary state, which can have a sacred aura about it and involve a state of *communitas* with others who are going through the same process. Speaking of his own experience in Mecca during the *Hajj*, Ahmad (1985:61) provides a religious example of this when he refers to the self-abandon of individuals and the great engulfing of the individual in the crowd. For Turner, this is a state in which the artificialities of daily life give way to a more spontaneous and liberated experience. In contrast to much of day-to-day life, it now becomes possible for usually unexpressed aspects of the personality to appear, sometimes in a memorable fashion. The rite ends with the process of 'reintegration' in which people pass back into the structured daily life of their society. This, itself, may require significant adjustments. Speaking of his return from the *Hajj* to Tehran, during which he had experienced many difficulties, Ahmad (1985:134) says, "I was home by five, frayed, coughing and exhausted. I haven't yet gone out, not even to the bathhouse, out of concern for the injury on my foot and for fear of catching cold again".

As an aside, it should be said that, in the Turner model, because the individual often is returning to another social niche after the ritual, as for example from child to adult or unmarried to married, the state of the society to which an individual returns is not (for him or her) the same as that from which they began, but it has a structured ordinariness, nevertheless.

For Victor and Edith Turner (Turner and Turner 1978:34–35), pilgrimage has the potential for producing, through travel, some of the essential qualities of *liminas*, but they think that it is a voluntary "not an obligatory social mechanism to mark the transition of an individual or group from one state or status to another", which is to be found in the transition rituals of tribal societies. The Turners (1978:253–254) use the term *liminoid* to refer to the less obligatory, more voluntary social and psychological aspects of pilgrimage. The notion of release from some routinized, social structure, however, is in their view, central to both pilgrimage and rites of passage at home. One should keep in mind, though, that in their view such a "release" does not free the individual from social constraints.

The Turner approach to ritual has attracted a number of anthropologists engaged in the study of tourism. It seems to have had some influence on Jafari's development of the springboard metaphor and on Graburn's profane–sacred–profane tourist sequence. Some people might have a problem of equating the playful or ludic "high" which seems to be the goal of a good many contemporary tourists, with the more serious business of religious ritual, but the notion that both tourists and pilgrims encounter some kind of *liminas* or *communitas* does have a surface plausibility. It may be that we have friends who have returned from vacations positively glowing from experiences in which the structured necessities of daily life at home had been sloughed off in favor of a more spontaneous, liberated existence. An experience recorded in the journal of the author in the Italian (Piedmont) city of Cuneo points to such an existence.

> Feeling good after having had an excellent dinner and some good Arneis wine, I was strolling along one of the arcaded sidewalks near Cuneo's central square. Suddenly, I heard singing, which turned out to be from a church service inside. Going in, I found a congregation standing and singing with hands joined. In their give and take with the clergy up front, they were responding so joyfully that I was 'thrilled'. Soon, the service was over, and I was 'thrilled' again to see people shaking hands and to have my hand shaken. For me, there came a wonderful feeling of belonging, which may have been particularly striking because of my being alone in a foreign place. On walking back to my hotel, I took special pleasure in noticing couples and families enjoying their togetherness. Grand! The thrill did not persist through the night, but I carry the memory of it as a treasured souvenir of my visit to that little city in the Italian Piedmont.

The Turner paradigm certainly would seem to apply to some tourist experiences and reactions. Lett (1983:47), who applies it to carousing charter yacht tourists in the Caribbean, says, "The 'vacation' enjoyed by the charter yacht tourists is a temporary interlude in the tourists' lives. Like the liminal stage of a rite of passage which lies 'betwixt and between' the recognized statuses of structured life, liminoid activities fall between periods of routine, ordinary activities". Wagner (1977) finds the Turner metaphors applicable to Swedish vacationers in the Gambia who, in contrast to their more formal behavior at home, abandon status distinctions and begin to call each other by their first names. Passariello (1983) sees it at work among Mexican vacationers at the beach who ignore ordinary food rules, overindulge, and leave their favorite restaurant in disarray.

So impressed with this view of tourist experiences is Graburn (1983a), that he suggests that they involve a process of inversion resulting from a need that all societies provide for in one way or another. So, the "collective effervescence" of the Australian aborigines that was associated with their visit to sacred sites (see Durkheim 1947[1915]) may be seen to derive from such a need. For Graburn, tourism provides an opportunity for experiencing a necessary change from the ordinary routine.

What Turner and those who have made use of his views seem to be getting at with their notions of *liminas* and *communitas* is an existential state which has been associated with a variety of personal transitions in individual life courses, a state which self-reflective anthropologists might attribute to their own field experience. For people in such a condition, the structured certainties of daily experience have given way to ambiguity or uncertainty. Then, they no longer have such a good idea of where they stand and who they are. Turner (1969:95), at one point, says that liminality is "frequently likened to death, to being in the womb, to invisibility, to darkness, to bisexuality, to the wilderness, and to an eclipse of sun and moon".

According to those anthropologists who have been taken with the Turner metaphors as they apply to tourists, the experience of liminality is associated with being carried away by normally unexpressed inclinations. Thus, according to Jafari (1987:153), this (touristic) mode "actually invites anti-structural manifestations and tourism assumes the qualities of play". These anthropologists may be aware that this kind of thing may happen at home – indeed, is supposed to happen at home – from time to time if, as Norbeck (1971:51) argues, human life is to be maintained. In this view, people everywhere need relief from the structure of social obligations and an opportunity to renew themselves, as seems evident in Japanese drinking parties or the coffee breaks of Americans working on assembly lines.

Tourism, it might be suggested, is one of the more efficacious ways of accomplishing these ends – more efficacious because the travel involved in tourism removes one physically from the home routine, and so, tends to eliminate that routine as a distraction. Thus, Graburn (1989a:36) says that tourism is one way in which humans' need for recreation (re-creation) is fulfilled, the kind of view that, according to Rojek (1989:73), is consistent with the humanist strain of thought in Western theories about leisure and recreation, in which the realization of the existential self in a condition of comparative freedom is stressed. Questions about this alleged need will be dealt with more fully later in this chapter and in the chapter which follows.

The Contact Situation and Tourist Responses

There are a number of problems with the view of tourism as a kind of transition ritual, including the difficulties in trying to pin down the meanings of its concepts, as well as the problem of identifying an alleged need for alternation or inversion of experience, which is assumed to underlie tourist and other leisure experiences. However, it also has some virtues. Let us agree that all travel-visiting has the potential for generat-

ing, through strangerhood, something like the experiential state of liminas, which seems to be not unlike the experience of anomie discussed by
Schuetz in his papers on *The Stranger and The Homecomer*, mentioned
earlier. Depending on their own nature and specific nature of their
strangerhood, people will experience and respond to it in various ways. As
pointed out above, anthropologists struck by the Turner metaphors have
often emphasized the transcendent quality of the touristic experience –
something like what happened to the author upon arriving at the old
fortified French city of Carcasonne for the first time or a friend's 'almost
spiritual' response to looking down into a very deep canyon in the
American southwest.

Under such circumstances, tourism can be seen as an activity that
seems to be capable of generating some kind of extraordinary experience
that may be hard to duplicate at home. Such a highly involving psychological state has been referred to by Maslow (1968) as a "peak" experience
and by Csikszentmihalyi (1975) as "flow". In the study of the festival of
San Fermin in Pamplona by Leimroth and Stevens (1984:57), an informant who has participated in the festival provides an example of this
when he says, "The entire week in Pamplona is lived on an emotional
level unequaled in intensity and completeness by any other social activity
I have known, including combat".

Should this be the ideal–typical model for looking at tourist experiences? Probably not, as some of those who have used the Turner metaphors in their research will admit. Lett (personal communication), for
example, whose use of Turner's concepts in his study of Caribbean charter yacht tourism was mentioned earlier, says that "no anthropological
account of tourism would be complete if it dealt simply with the liminoid
experiences of the tourists". Nevertheless, the anthropological literature
on the subject of tourist experiences has been dominated by the Turner
scheme, and there appears to be a disturbing assumption among some
adherents that a tendency towards anti-structural manifestations while
touring does, in fact, reflect some universal human need.

To summarize the ritualistic view of tourism transitions, consider the
position of Graburn in his seminal pieces on the subject (Graburn 1977,
1983a). Graburn's argument here is that touring provides an opportunity
for humans to satisfy a need for variation from routine experiences. The
experience and reactions of a person while touring, which take on a
sacred aura and function to recreate the individual for day-to-day life at
home, are dictated by the need for inversion as well as by certain social
influences.

This view of tourism as a personal transition, which lays out the ritual-
like nature of the tourist's experience and reactions, as well as their
cause(s) and consequences, would seem to orient research more towards
the need which tourism is supposed to gratify rather than the social

influences which shape it. But the existence of some human need of this kind has so far not been adequately demonstrated. Why not, then, put it aside or in brackets and get on with what can be demonstrated, namely the variety of tourist experiences and reactions, their social causes and consequences? This will be the line taken in the rest of this chapter.

In exploring the variety of tourist experiences, which Cohen (1979b), among others, has stressed, consider tourism's negative side. Graburn has tended to emphasize the sacred "high" of tourist experiences. Certainly, some tourists offer evidence of this. But Sutton (1967) was one of the first to point to the negative aspects of the tourist's experience, and though he feels that the shorter visit and narrower range of tourist contacts render tourists less susceptible to psychological disturbances, Furnham (1984) has raised the specter of culture shock among them. Added to this, one ought not to forget the various hassles associated with travel which may leave some tourists feeling not so good. Gaulis and Creux (1975:11) say, "Pendant les siècles, le voyage n'etait qu'une suite de contretemps". Consider further one of the entries in Byron's travel journal, *The Road to Oxiana* (1982:286). He is comparing his return trip to England by P and O passenger steamer to an earlier one:

> I detect a change for the better in manners and obligingness. Also, the boat is only half-full so that we escape the communal life of a boarding-house. None the less it is an appalling penalty: a fortnight blotted out of one's life at great expense.

Or consider one of the author's entries in his journal, written while he was stuck in a Cubatour bus at the airport in Havana during a visit to revolutionary Cuba:

> The problem, says the guide, is that one person is missing... One American says that we should take up a collection to send the missing person into Havana by taxi. Others say to just leave him; we are wasting time. The complaints grow nastier and louder. Finally, someone suggests that we all take taxis. 'There aren't any', says the guide. People are very disgruntled. The plane arrived at 3. It is now 5, and still we are waiting.

The negative–positive dimension is only one aspect of the tourist experience. Cohen sees tourists clustering into five modes of experience. His "recreational mode" and "diversionary mode", which may be predominant in the contemporary world, seem to be far removed experientially from his "existential mode", which most closely approximates what Turner's followers seem to have in mind. There is also an increasing array of subjective reports by tourists, which can be ordered in terms of Cohen's or some other scheme.

The work of psychologists and other social scientists, who have been probing into the phenomenological side of human actions (see Mannell and Iso-Ahola 1987:325–327) can be applied to tourism. As an example of this kind of data, take the observations made by the author in his journal during a visit to the northern Italian city of Torino.

> I saunter down to the Po river. I like seeing the waiters carrying trays of coffee to nearby offices and the relaxed pace of the traffic (keep in mind, though, the terrible revolutionary slogans painted on the wall and the horrible work of the Red Brigades in this town). As I saunter along, I, miraculously, begin to relax. What has caused this? Is it the city, the fact that I've taken care of my reservations, or that I'm just beginning to feel at home? Whatever. It's a beautiful city and I'm relaxing into it.

Such reports are now being obtained by more advanced techniques including the bleeper–recorder (see van der Wurff *et al.* 1988), which promotes more systematic observation by signaling subjects to record information directly onto the portable recorder they are carrying. More reports, obtained by more methodical procedures, are needed to add to the fund of tourist experiences on which this aspect of tourism research should be based.

Aside from the alleged need for alternation or inversion, which has been put aside or in brackets here because of problems with demonstrability, there are a number of factors, only a few of which are referred to by Graburn, that could assume prominence in shaping tourists' experiences and actions. First, consider the kind of tourism being practiced, that is, the role tourists play during their sojourns. Is it mountaineering, sailing, lying on the beach, group sightseeing, museum-going, skiing, taking the waters at some spa, or whatever, that occupies the tourist? Yiannakis and Gibson (1992) have proposed 13 different tourist roles and associated experience modes for the group of Americans they studied, and they go on to suggest a number of factors which cause tourists to "select and enact" these roles.

Second, is the tourist alone or in a group? Here, the Turner concept of *communitas* is ready at hand for describing the experience and actions of tourists. If tourists are traveling with others, these others can minimize encounters with the hosts and their culture, and so reduce contact with the *liminas*-laden foreigner. Peer groups of tourist-strangers sometimes can dominate tourists' lives and regulate their actions. Bochner (1977:122) and his colleagues, for example, have pointed out that overseas peer groups can be of vital importance in the adaptation of student strangers; and Herman and Schild (1960:236) found that for a group of American students in Israel the stranger group "played a crucial part in the development and maintenance of students' attitudes".

The Turner-related ethnographic work with such tourist groups by Lett, Passariello, and Wagner already has been mentioned. An additional interesting ethnography of such a group, which its author refers to as a short-lived, little community, was carried out by Foster (1986:228) during a cruise in the southwest Pacific on a small cruise ship. Foster states that "the essential equality and easy interaction among passengers... contributed to the pleasure of the trip... and to the rapid welding of the passengers into a group that saw itself as cohesive and relaxed". Considering the importance of stranger groups in the tourist odyssey, many more ethnographies of them are needed.

An important research question (which obviously has an applied side from a marketing point of view) concerns the solidarity or morale of such groups. Nash and Tarr (1976:367–368), studying a group of American students in France, point out that "the most evident consequence of their low morale or cohesiveness... appears to have been a reduction of the significance or meaningfulness of the total overseas experience". And Quiroga (1990:199), on the basis of a study of Latin American tourists on guided coach tours in Europe, reports that "the way in which tourists evaluate shared group experience has a significant influence on their final satisfaction".

Not only tour groups, but their guides have been implicated in the reaction of tourists to their journeying. A special issue of the *Annals of Tourism Research* (Cohen 1985), which explored the role of tour guide, prepared us for findings such as those of Geva and Goldman (1991:180), who, in their study of Israeli tourists traveling to Europe and the United States, found that the guide was the single most important factor in tour satisfaction. As Quiroga's (1990:199–200) data show, a guide can promote satisfaction, first of all, by fostering social interaction, which was seen by her respondents to be the most important function of the tour guide.

One thinks back to the tutors of those young English gentlemen on the 'Grand Tour' who were, in fact, a kind of guide. What kind of relationship did they have with their young charges, and how did it contribute to their overall education? Considering the amount of historic material available on the Grand Tour (see Towner 1985, 1988), the exploration of this historic question would seem to be as worthwhile a project as that concerning present-day students abroad, for whom, as Nash and Tarr (1976:369) point out in their study, the quality of leadership (by faculty and student advisors) of their overseas study group seems to have been associated with group morale.

Another factor shaping tourist experiences is the nature of the host society and the relation of the tourist to it. Here, the work of Simmel (1950), who explored the social position of the stranger in some detail, would seem to be a particularly useful heuristic device. Tourist friends often report on the kind of welcome they received from their hosts. Were they greeted with courtesy? Did people on the street go out of their way to give directions? Were they, somehow or other, made to feel at home?

Studies of adaptation to another culture have shown that strangers tend to get along better if the relevant host culture is more compatible with theirs. Simmel, in one of his provocative observations, says that the stranger who is near (in a physical sense) also is far (in terms of cultural distance). But some strangers are less far, culturally, from their hosts than others. Furnham and Bochner (1986) report on a series of studies that link the principle of compatibility of cultures with individual adjustment. So, other things being equal, mainstream American tourists should tend to get along better with, say, Norwegian hosts than with Rumanians. Further, if a stranger is able to work out some kind of cooperative relationship with hosts, the going tends to get easier (see Brislin, 1981: 184–185). A tourist interested in archeology, for example might seek out some local counterparts to see if he or she could help out in some dig. Finally, welcoming overtures by hosts seem to help strangers feel more at home. Nash and Heiss (1967:220–221) found that, for their laboratory strangers, the more accepting their hosts the less likely they were to experience stranger anxiety.

All of this, of course, has to be seen not only in a framework of the specific cultures involved, but also in the context of the relationship between the societies in tourist–host contact. If all goes well, as in Grenada after the American invasion, American tourists are likely to be welcomed. If a country is very powerful, as Great Britain was at the turn of the century, its tourists will share in that power, and the burden of adaptation will fall more heavily on the hosts. Thus, when the power of Great Britain was supreme around the world, English tourists found it easy to have their way not only in places within the empire, like India or Rhodesia, but elsewhere such as St Moritz or the French Riviera. The not-so-facetious title of a book by Howarth (1977), *When The Riviera Was Ours*, alludes to this. And one should not forget the relevance of studies of prejudice against immigrants which have, over the years, indicated that the attitude of hosts towards such strangers is dependent on, among other factors, their numbers and rate of arrival.

The nature of the home society and the relationship of the stranger to it also could have an effect on the tourist's experience and reaction. Nash and Heiss (1967) also found that the greater the commitment of laboratory strangers to their home group the more likely they were to experience stranger anxiety. The home culture also can shape tourist actions. Passariello (1983:121) offers a dramatic example of this in her study of middle-class Mexicans at a small beach resort. Contrasting the excesses of her Mexican vacationers with the more reserved behavior of foreign visitors, both of whose actions might be construed as supporting some kind of inversion principle, she (Passariello) says that though the motivation, symbols and reality adjustments of the two groups may be similar, "the structure and form of their touristic experiences vary culturally and historically".

To conclude this discussion of some of the more important factors shaping tourists' experiences and reactions, consider the characteristics of individual tourists, as did Ireland (1990) in an intriguing study of the experiences of visitors to Land's End in England. In this study, Ireland was able to show that the degree to which people were annoyed by the presence of other tourists during their visit was significantly related to such variables as occupation, age and gender of the visitor. Though Guthrie (1975, 1981) has argued that enduring personality attributes are less important than situational factors in the adaptation of strangers, it would be a mistake to entirely forget about such qualities in tourists. Furnham and Bochner (1986) and Nash (1970) have discussed some of these qualities and how they relate to adaptation abroad, the evidence for which often is inadequate.

One problem that besets research on the relationship between personality attributes of strangers and their adaptation abroad is the variability of the contact situation. Tourists who have to deal a lot with surly Parisien functionaries may have a harder time than those who manage to by-pass them. So, too, would the tourists who stay in hotels where the personnel do not speak their language nor understand their ways. Also, tourists traveling in groups would experience different adaptive problems than those touring alone. It may be that those in group tours may have even more problems relating to their fellow tourists than to the host population. All of these observations, of course, assume that we know what tourists are up to. As Cohen (1979b) points out, their inclinations at the outset will tend to condition all of their experiences and reactions. This suggests that any use of the paradigm of tourism as a personal transition must begin with the tourist's initial inclinations at home, which is, indeed, suggested by Graburn's adaptation of Turner's theory to tourism research.

To sum up this discussion of the structure of the contact situation and tourist responses to it, it may be helpful to return to the Turner concepts that have excited anthropological interest. With the benefit of hindsight, it now seems that they have had only limited usefulness for the advancement of research on the subject. The main problems seem to be the difficulty of operationalizing them and the assumptions of some of those using them that there is a universal need for alternation or inversion of life experience, which tourism (as well as other leisure activities) can satisfy. It may be, as Nash (1984:505) indicates, that there is some such need, but if there is, it certainly has not been adequately established. Consequently, the first order of business in looking into tourist experiences should not be to assume such alternation. By concentrating on such an alleged need and the ways of satisfying it, researchers may be prone to miss or de-emphasize the tremendous variety of tourist experiences and the circumstances that shape them.

On the other hand, the Turner paradigm, as employed by anthropologists studying tourism, has performed a salutary service by causing researchers to look at tourism as a personal transition and explore the phenomonology of tourist experiences. Also, the concept of *communitas*, as it relates to the morale or cohesiveness of tour groups, has been useful in pointing to the role of peer groups in shaping the experience and reaction of tourist strangers. So, one can see that this anthropological theory has been useful in sensitizing anthropological researchers to certain aspects of their subject matter. However, if too narrowly employed, it would not seem to provide a particularly productive line of research for the future.

One worries about the attractiveness of Turner's views, particularly to researchers in Western societies where the realization of self has become such an important issue. It could divert attention from the tremendous variety of tourist experiences, some of them rather mundane, and the circumstances that shape them. The first order of priority for anthropologists should be to explore in depth the entire range of tourist experiences and associated activities and attempt to classify them in some theoretically viable way. Here, the work of Cohen (1979b, 1988b) and Yiannakis and Gibson (1992), among others, can serve as a useful point of reference. Then, referring to the array of factors just discussed, one might begin to attempt some explanations.

Effects of Touring on the Tourist

How much and in what ways are tourists affected by their experiences abroad? This question follows logically from the conception of tourism as a personal transition. Csikszentmihalyi (1981) argues that leisure experiences (of which tourism is one) can provide fertile ground for developing enduring standards of behavior. According to him, the expressive activities of leisure become the basis for evaluating all of one's other experiences. For those acquainted with the cultures of North American Indians, this may call to mind those quests by young people for supernaturally-given visions that would sustain them throughout their lives.

But only a little acquaintance with present-day tourisms (e.g., Cohen's "recreational" or "diversionary" modes) should make one skeptical about their power to affect individual tourists. Ideas that tourism will promote a broader, more tolerant world view or a more relaxed way of living, for example, ought to be taken with a large grain of salt. Too many tours are too short, too superficial and have qualities too much like home to result in enduring personal transformations. There may be some tourists somewhere who carry with them residues of some profound tourist experience. Perhaps some of them have become patrons of those Sherpas studied by Vicanne Adams, whose work was mentioned in *chapter 2.* But the skeptic would think that such people would not be very numerous among the legions of contemporary tourists. All of this is speculation, however. It would be better to refer to studies of the subject. What do they tell us about the effects of touring on the tourist?

To begin, consider some methodological problems. The kind of study that will actually find out what touring does to tourists is not easily accomplished. Putting aside, for the moment, all those not so insignificant problems of getting at elements of an individual's character or personality, there is the matter of discovering the changes that are, in fact, brought about by tourism. It will not do to simply find out whether the individual has changed since departure. That change might have come about because of factors other than tourism; or it might be simply the result of being asked the same questions over again by the researcher. In order to rule out such factors, there needs to be a control group which is, as much as possible, like the tourists. This group would remain at home, and like the tourists, be assessed before, possibly during and after the tour; and in order to see if any changes in the tourists are enduring, it would be necessary for subjects to be available for assessment for some time afterward. Church (1982) has discussed the methodological issues in the investigation of sojourners, generally, and Philip Pearce (1982a: 86–93) has considered them for the study of tourists, specifically.

What are the kinds of personal reactions and longer term effects that should be of interest to an anthropologist? First, consider the matter of cross-cultural adaptation, with which many anthropologists have become intimately acquainted during the course of fieldwork. How well do tourists adjust during their tour and on their return? Does the encounter with the new and different raise problems? If so, how quickly are they resolved? If tourists have been troubled during their touring, they are more likely to trouble their hosts and their travel agents afterwards. Such troubles also reveal interesting details about the psychological side of culture-contact and change.

A persistent notion in sojourner research is that there is a typical course of sojourner adjustment, which conforms to a U- or a W-curve. Thus, people are supposed to experience a heightened sense of well-being at the outset of their foreign encounter, then a decline as the frustrations of operating in a different world mount up, then increase again as the new and different is mastered (the U-curve). The same course is supposed to be followed (making for a W-curve overall) on returning home to a situation that turns out to be different than expected. Unfortunately, support for this hypothesis, according to Church (1982:542), "must be considered weak, inconclusive, and overgeneralized".

In one of the more adequate studies of this issue, Nash (1991), following the course of well-being of an overseas study group in France and a control group at home, found no support for the U-curve hypothesis.

Evidence concerning some universal course of tourist adjustments is even less convincing, though a study by Philip Pearce (1982a:41–46) of tourists on one-week vacations to two Australian islands does suggest something like a U-curve of adjustment, with negative, health-related moods more likely to be reported in diaries during the first few days.

Many more methodologically adequate studies of the course of adjustment of tourists certainly are needed. Also needed is an awareness of the kind of tourism being practiced and the context in which it takes place. A number of significant others figure in the tourist's odyssey. These include host and home groups and often a group of fellow travelers, any or all of which can exercise a significant influence on the tourist's experience and behavior and leave their mark on his or her character. Questions about the broadening effects of tourism, for example, have to do with the hold of the home reference group over the tourist during and after the tour.

Touring takes a person out of some daily routine that has the weight of a society and its history behind it into a world of leisured travel. Any psychological changes associated with this change of situations should be measured from the base line of the home society and its routine in which the individual develops a view of self and others (animate and inanimate) that is so reinforced that it becomes, in Schuetz' famous words, "a world taken-for-granted". Moving out of that situation, for any reason, and entering the *liminas*-laden world of travel ought to undermine the hold of this world on the individual. At the very least, the changes brought about by tourism or any other form of travel might be expected to produce less ethnocentrism. Significant figures who might be perceived differently as a result of touring are the hosts, who may have figured only a little in the tourist's world view prior to departure; the home society, which has come to be taken for granted; and the self, which is intimately linked with the significant others in an individual's world view. Research into the personal consequences of touring ought to take changes in the tourist's perception of these important figures as a priority.

A number of studies of attitude change as a result of touring have been carried out. Pearce (1982b), for example, tried to find out if post-travel, tourism-related images of host societies differed from pre-travel images. Using three groups of British subjects, one traveling to Greece, another to Morocco and a third (the control group) which remained at home, Pearce asked respondents to indicate how much they agreed that a country offered such things as cheap shopping, adventure, good sun and beaches, or a swinging social life. He found that, compared with the control group, both groups of travelers changed some of their images of visited locations in a positive direction. In addition, they saw their country of origin (Great Britain) differently upon their return. Finally, in a demonstration of a 'generalization effect', there was a spread of changes in perception of destination areas to similar environments such as Tunisia (for Morocco) and Italy (for Greece). This study demonstrates that, not only can changes in attitudes toward the various significant others involved in one's touristic odyssey occur, but that these changes are interdependent. No attempt, however, was made by Pearce to see if these changes persisted.

As far as changes brought about by tourism on the nature of the self are concerned, Brown's (1992) ingenious study of respondents' associations to personal possessions acquired on tour, such as tour photographs, suggests how an individual's self-identity can become linked with "touristic life, destinations and social roles". More systematic evidence is provided in a study by Nash (1976) that used a group of American students in France and a control group at home. This study shows that, after their return from a year in France, the overseas group showed a significant increase in autonomy of self, an expansion and increased differentiation of self (more self conceptions reported) and a decrease in alienation from their feelings and from their bodies. An attempt was made to see if these changes endured by testing the students several months after their return. Though hampered by somewhat inadequate data, this study indicates that, for the most part, the changes noted immediately after return home did not persist in the students who had been abroad for the academic year.

So much for the effects of single trips abroad on tourists. The meager data available do not make one optimistic about the persistence of any changes; nor can we be encouraged in such a view by Graburn's confident assertion that tourism recreates people and hence is a device for putting hard-pressed people (and their society) back in order (see also Krippendorf 1986:524–525).

But what about a possible build-up of touristic effects from a long-term visit or as a result of repeated visits to the same, or similar locations, as with vacationers from the French city of Bordeaux to their nearby retreat, Arcachon, the Japanese to their favorite spas, the colonial British in India and other outposts in southeast Asia who escaped the bad season by going to villages like Simla in the hills, or those French Canadians who pass their winters in Miami Beach, Florida year after year? Do such tourists carry with them enduring marks of these visits? What of the Piaroa youths of Venezuela, who, according to Kaplan (1975) regularly roamed together round the widely dispersed Piaroa homesteads? And what about all those young men from England in the 1920s for whom, according to Fussell (1982:vi), "travel became an obsession?" Does their cumulative travel and the travels of other 'world travelers' indeed, leave the kind of residue that is suggested by the old saying that "travel broadens the mind?"

Obviously, much more research is needed before we can accept with any confidence the notion that touring of whatever kind has consequences of significance for the tourist. While waiting for such research to be done, it might be useful to pay attention to tantalizing personal observations such as those of a prince of travel writers, Robert Byron (1982:280–281), who refers to a stop on the road back to Kabul as follows:

> Some small trees of the sallow type were growing along a stream near the road, and Seyid Jemal stopped to let his assistant pick a few branches from them which he threw into the back of the lorry. As they fell at our feet they gave off the same elusive smell which has pervaded the whole journey since we first met it at the Afghan frontier. It emanated from clusters of small green flowers which are unnoticeable from a distance, but which, if ever I smell them again, will remind me of Afghanistan as a cedar wardrobe reminds me of childhood.

Changes in Attitudes of Greek Tourists Visiting Turkey

Petros Anastasopoulos is an economist who has been broadly trained in that field. His interest in tourism seems to be related to a specialty in international business. His economic expertise, which include statistical techniques, with which most anthropologists have not acquired a working relationship, has been brought to bear on a variety of tourism issues. A Greek background has helped him in several studies of Greek tourism and tourists.

His study, which will be considered here (Anastasopoulos 1992), did not draw particularly on his expertise in economics. It was one project in a broader investigation into the effect of touristic contact on tourists traveling between traditionally hostile countries, as for example, Israeli tourists to Egypt. The study was guided by the broad hypothesis that such contact would result in changes in attitudes towards the destination country and its people. In addition, the changes were thought to occur in a positive direction. Unlike the Pearce study, mentioned earlier, specifically touristic items were not included in the questionnaire. Rather, tourists were asked to evaluate only host peoples and their culture.

The Anastasopoulos study involved the use of 97 first-time Greek travelers on six one-week bus tours to Turkey and a control group of 82 at home. Attitudes of these two groups about Turkey and its people were assessed before the tourists' departure and after their return. The author raises some questions about his questionnaire, versions of which already had been used by others in previous studies in the same project (Americans to the Soviet Union and Israelis to Egypt). In doing this, he helps us to better understand the results. He also points out that his tourist group and control group were not perfectly matched (a perennial problem in studies of this kind), with the touring group tending to be more favorably disposed to the host country at the outset. However, it does not appear likely that either of these methodological problems had a significant effect on the general pattern of attitude change which was found.

By comparing changes in the evaluation of the Turks and their culture between tourists and control group, the author was able to assess the impact of the tour on the Greek tourists. Using the appropriate statistical technique (in this case, a *t*-test), the author found that, as compared with the control group, the tourists' attitudes generally shifted in a statistically significant *negative* direction. With exceptions for only a few items, Turkish institutions, quality of life and the character of the Turkish people tended to be viewed more unfavorably after the tour. The rosy-hued hypothesis that there would be a positive change in attitudes, therefore, was not confirmed.

Why the negative attitude change among the tourists? The author, who clearly expected that his positively-oriented hypothesis *would not* be supported, accounts for this in terms of the conditions of contact of the tourists with their hosts. For example, he points out that in contact studies a favorable social climate has been associated with a positive attitude change, but during the study the Turks were continuing to anger the Greeks by their actions in Cyprus. Contact studies also have suggested that intimate social contact is associated with more favorable attitudes about the other, but as Laxon (1991:373) also has pointed out, superficial touristic contacts, such as those that existed during the Greek bus tours, would not be likely to have this effect. Anastasopoulos cites a number of other social factors that could have affected his tourists' attitudes, but one misses the kind of down-to-earth analysis of the social conditions of these tourists' contact that is normally carried out by an ethnographer. The author apparently recognizes this, and he does suggest what aspects of the acculturation situation of 'culturally diverse tourists' would have to be rearranged to ensure a more favorable outcome.

This study by Anastasopoulos of the effects of a bus tour to Turkey on Greek tourists is theoretically informed by a contact hypothesis that was developed in the United States for studying relations between ethnic groups in, for example the integration of public housing. But factors that may have proved favorable to integration in American culture might not operate in the same way elsewhere. For example, according to one version of the hypothesis, equal status contacts are supposed to promote more positive attitudes and better relations among the parties involved. This may be so in a culture that, like that of the United States, has an ideal of equality, but there is a question whether it would have a favorable effect in societies where status differences are more acceptable. Hopefully, the earlier discussion in this chapter of factors affecting the experience and response of tourists has prepared readers for a more critical reading of the Anastasopoulos article.

On the methodological side, the study does make a laudable use of a control group, which permits the author to say with some certainty that the tour was responsible for the mostly negative changes in attitudes of the Greek travelers towards their hosts. It is important that other studies of this kind pay more attention to the acculturative context and to the specific roles played by the tourists in it so that it will be possible to find out what it is, exactly, about a tour that is responsible for the changes in attitudes of the tourists involved. Of course, a follow-up on the views of the Greek travelers, as well as the controls, over a longer period of time would have been desirable in order to assess the persistence of the attitude changes, but this longer-range project still seems to be a utopian dream for those having an interest in this kind of research.

At the very least, this interesting study by Anastasopoulos demonstrates how dangerous it is to make generalizations about all of tourism, however conceived. Not only should it now be evident that tourists' intentions and actions vary considerably, but tourism's consequences, whether for hosts or the tourists, themselves, also vary. In discussing the effects of various kinds of contact situations on tourists, Anastasopoulos adds another voice to those, including Cohen, who have been calling for a recognition of tourism's diversity (see also Smith and Eadington 1992). His research also should provide a useful steppingstone for future investigations of the subject.

* * *

The effects of touring on the tourist also can be assessed as more or less satisfying, an issue in which the travel industry has an important stake. If tourists are not satisfied with their trips, their dissatisfaction will ultimately show up on the 'bottom line' of those who have a commercial interest in tourist ventures abroad. It is not surprising that a good deal of market research has been conducted to assess tourist satisfaction. Typically, tourists are asked about their attitudes towards a particular tourist destination before and after their trip (see e.g., Pizam *et al.* 1978). This can be put to anthropological use by relating tourist satisfactions to the roles they have played and their intentions at the outset, that is, by looking at the transition process in its entirety. As Mannell and Iso-Ahola (1987:321–325) have pointed out, tourist motivation is an important dimension of tourism research, but methodologically adequate studies are practically non-existent.

One also might consider tourist motivations as part of some social agenda. An example of this is the adoption of a policy by the Japanese Government of promoting two-way international tourism (JNTO 1992:79–80) which seems to have resulted on the Japanese side in an increase in Japanese tourists going abroad. Another is the promotion of youth tourism by the Catholic church in Poland as one aspect of a desired socialization process by that organization (see Swianiewicz and Swianiewicz 1989). Are disaffected workers actually restored by a paid vacation at the beach or in the mountains as the social planners of this practice envisaged? Individual tourists may not know that they are acting out the tourism program of some institution or society, but any anthropologist should be aware of what that agenda is and how well the needs or motives which it played a part in developing in the first place are being satisfied. Here, Rojek's (1985) discussion of the structural characteristics of modern leisure practice and of various sociological theories for dealing with it have relevance for anthropological analyses of the consequences of touring for tourists and their societies.

And this is not all. The home society may not be the only significant other to have a stake in the effects of touring on tourists and their satisfaction afterwards. Hosts and those who carry the tourists here and there (airlines, railroads, rent-a-car agencies, etc.) also should be interested. Indeed, the tourist may figure in the agendas of a variety of parties involved in his or her personal transition, each of which depends on the satisfactions of individual tourists for its success. Any assessment of the consequences of touring for the tourist and their societies, therefore, would do well to keep all of these interested parties in mind. This, of course, should not be news to the anthropologist, for whom the issues of human social life, including touristic transitions, are routinely considered in their relevant social context.

Finally, it might be wise to get back to the issue of tourism as a personal transition. As was suggested in the important work of Turner (as adapted for the study of tourism by Graburn), this perspective involves more than the study of the experience and reactions of tourists. It also is concerned with what motivates tourists in the first place and the consequences for them and their home society. As should be clear from all of the preceding, the field is open for investigations of all of these issues. The ritualistic notion of tourism, a laudable and unusual thrust in anthropological thinking about tourism, has served to inspire such research, but, as should be evident, it would seem to carry us only a little way in such a program of study.

Chapter 4

Tourism as a Kind of Superstructure

In earlier chapters, it was pointed out that most of those who have been studying tourism from an anthropological point of view have looked at tourism from the perspective of acculturation or development, and in so doing, have concentrated on its *effects*, particularly on host peoples in less developed societies. A few, looking at tourism as a personal transition, have taken the experiences of tourists, their causes and personal consequences as central to their investigations. Even less attention has been given to the ultimate causes of tourism, that is, the conditions (on what economists refer to as the demand side) that generate tourists and tourism. Thus, what usually is thought to be the main concern of any science, which is the search for causal explanations of things (see Kitcher 1993:127–177), has so far been largely neglected.

It may be argued that tourism has been seen as a cause of social and psychological changes in hosts and tourists, but that leaves aside the question of what brought about the tourism in the first place. In this chapter, an attempt will be made to repair that deficiency by gathering together whatever anthropologically-oriented work on the subject that does exist and laying out a broad conceptual scheme for handling questions about the cause or causes of tourism.

How to account for the existence of tourism or different forms of tourism? As far as the touristic process outlined in *chapter 1* is concerned, the answers to such questions are to be found in the tourist generating situation, in which the home society certainly plays a crucial role. With anthropology's traditional preoccupation with the less developed world, home societies of anthropologists, mainly those in Europe and North America, have received comparatively little attention. Yet it is in such home societies where many of the root causes of tourism are likely to be found.

As pointed out in *chapter 1*, anthropologists are showing increasing interest in the study of their own societies, which gives some hope for the investigation of questions concerning the generation of tourism. A number of anthropologically-oriented scholars have issued calls for the study of this aspect of the touristic process. For example, Hamilton Smith (1987:343) speaks of the need to study the relationship between tourism behavior and home behavior; Graburn (1983a:19) argues that we should seek to explain "why specific touristic modes are attached to particular social groups at the historical period when they are found"; Nash (1981:465) says that the study of the tourist generating situation is "indispensable" for understanding the entire touristic process; and Crick (1989), in an influential article in the *Annual Reviews*, has stressed the importance of such research.

Heeding these calls, researchers could investigate such questions as why people of higher rank tend to engage in more protracted, longer range, more luxurious forms of tourism (see Nash 1979b:22), why the Japanese prefer to tour in large groups (see Graburn 1983b), or why the percentage of Norwegians 'not' taking a vacation has tended to decline in recent years (see Haukeland 1990:174–175). The general questions on this agenda can be phrased as: 1. Why tourism? 2. Why differences in rates and kinds of tourism? In keeping with the anthropological mission, these questions would be asked for the entire range of human societies where tourism has occurred.

To begin to answer such questions, recall how sociocultural anthropologists have tended to explain human social actions. For example, to account for the general human tendency to avoid mating with close relatives (i.e., incest avoidance), they have referred to either in-born (biological) mechanisms or some universal features of human society. Cross-cultural variability in incest avoidance (why, for example different peoples tend to avoid certain relatives), has been explained overwhelmingly by reference to sociocultural factors. In looking at such factors as influences on social action, anthropologists usually have begun with the holistic view that different aspects of a society's culture (e.g., its economy, religion, language) and its environment are related to one another. They may reinforce or conflict with each other, but they are seen to reveal some kind of interdependence, nevertheless.

Some social scientists believe that they have found a kind of causal 'Rosetta Stone' in one or another of society's institutions. For example, it might be some aspect of the economic system that is conceived to be ultimately determinate, as in Marxist analysis. Others are more eclectic in regard to which institution has the ultimate shaping power at a given time in its history, with now one, now another gaining dominance, as in the kind of analysis pioneered by Max Weber. As far as the home societies of tourists are concerned, it is very hard to think of any example where tourism has achieved dominance in this respect. Indeed, though tourism may have some influence in shaping other sectors of a home society, the evidence suggests that here, on balance, it tends everywhere to have a dependent nature. This is revealed when an economy sags and tourism monies and other parts of the discretionary income used for tourism usually are the first to dry up. Because of this, it seems appropriate to think of tourism mainly as caused by other aspects of the home society's culture rather than being the cause of them.

The Nature of Tourism and its Distribution in Human Society

Taking the tourist to be a leisured traveler, questions are raised about the nature of, and conditions for leisure and travel. Leisure is the key concept here. Though defining it continues to present difficulties (see Cooper 1989; Neulinger 1981; Kelly 1983), there is considerable agreement that, at a minimum, it involves *freedom from* social obligations that are essential for the maintenance of a society and its reproduction – obligations that Dumazedier (1968) refers to as "fundamental" or "primary". Many authors take these obligations to center around work, broadly or narrowly defined, which, because it is crucial for a society's functioning, is supposed to have a more socially compelling nature than other activities.

Leisure activities such as tourism, on the other hand, do not rate very highly as far as human survival is concerned. Looking at Suomi's (1982:169) analysis of play, which has been closely tied to leisure by some authors, one can understand why. He says, "There seems to be general agreement that the evolutionary costs of play are large and the immediate benefits small. The 'true' benefits are thought to accrue later in life, usually after play behavior has all but dropped out of the individual's repertoire".

People who are not engaged in work or work-related activities, that is, in fulfilling what Dumazedier's refers to as primary or fundamental social obligations, may be thought to enter the sphere of leisure time and then to realize their leisure in various socially structured ways such as lazing in the sun, gossiping with neighbors, or (in a form of tourism) vacationing in the mountains. According to this view, leisure activities are those which are performed in the time available "beyond existence and subsistence time" (Clawson 1964:1). Rojek (1985:85–105) takes issue with this view, but his criticisms can be accommodated by taking leisure to be a cross-cultural universal, the amount, form and allocation of which is culturally and intra-culturally variable.

Studies of time use by anthropologists provide support for these propositions (see Gross 1984 for a review). Johnson (1978), for example, looking at time use among the Machchiguenga, a horticultural people living in the tropical rainforest of Peru, divided up their days into production time, which includes all forms of work, consumption time, in which goods and services are consumed or used, and free time, which is the time left over from the first two. Though his conception of free time is more restricted than the notion of leisure time used here, he was able to figure out something like the amount of leisure time available to the Machchiguenga and compare it cross-culturally. For example, he found that, compared with a sample of French adults, the Machchiguenga had more 'free time'. Why this was so is the kind of question that will be asked here, not only for leisure, but for leisured travel, which is how tourism has been defined in this book.

The well-used *Outline of Cultural Materials* (Murdock *et al.* 1982), which charts various areas of human social life, considers leisure as one of the cultural universals, which, as mentioned earlier, sometimes have been explained in terms of some evolutionary-given attribute of human beings (see Brown 1991:39–53). As far as leisure is concerned, it has been suggested that humans have a need either for leisure itself (Sayers 1989:42–43) or for variation of experience around some optimal existential state (Iso-Ahola 1983), as when they go on vacation to escape an overload of sensory input in their day-to-day routine. As was pointed out earlier concerning Graburn's view of tourism as a personal transition, the existence of some kind of need for alternation or inversion of experience has not been satisfactorily demonstrated. The same can be said about any 'need' for leisure.

It is easier to suggest a 'need' for work or work-related activities in all societies (according to the *Outline of Cultural Materials*, all societies have work-involved economies) which are aimed at satisfying basic human wants. Going further, it can be shown from cross-cultural studies of time use, that in no society do people work all the time. They enter the realm of leisure in the time left over from work (minus necessary recuperative activities like sleep, rest, etc.). Of course, there is no clear line of demarcation between work and leisure. Perhaps it would be best to think of people as being more or less at leisure in their various activities. Thus, someone on vacation may call the office from time to time to add a little work to their leisurely schedule; and people at work may joke around, gossip, and engage in various forms of pure sociability.

All of this suggests that any attempt to define the leisure of a society will have to refer ultimately to the way in which its people work. So, Nash (1979a:4) has argued that leisure is a social production ultimately dependent on the nature of a society's working arrangements, which leave people not only more or less free to experience it, but also generate the surplus goods and services needed for people to enjoy it. As Sahlins (1972) has argued, there are two ways to produce a surplus. One is to produce more, which accounts for the fact that leisure time (for some peoples, at least) has tended to grow with industrialization. Those who have seen tourism to be a form of leisure, including the author in an early paper (Nash 1977:36–37) have taken this view. The other strategy to gain a surplus is to want less, which, as Sahlins points out, is the case with some wandering hunters and gatherers like the San of southern Africa or the Murngin of Australia, who, before they were forced to settle down, did not need to work so much to support an often abundant amount of leisure time.

Anthropological efforts to chart the distribution of leisure have gradually moved beyond the simplistic notion that there are 'primitives' in the world, like Johnson's Machchiguenga or those referred to by Sahlins, who have more leisure time than do people in more complex societies such as those in present-day Europe, Japan, or North America. Chick (1986:162) argues that the relationship between leisure time and social complexity takes the form of a U- or J-curve, with cultures at the low and high end of some complexity scale having the most and those in between (agricultural societies, mostly) having the least.

One has to be careful with such sweeping generalizations about the relationship between the amount of leisure time and social complexity or development, however. First, as Bird-David (1992) has made clear, there are so many problems with time allocation studies that any generalizations have to be made with caution. Second, to speak of free or leisure time in a global or per-capita sense may do violence to social reality. In no society does everyone have the same amount of leisure time, and in some, say, a typical agriculturally-based state, the differences between different social strata may be great. There, numerous peasants will have to work long and hard to support their rulers' leisure activities. In less stratified societies, generalizations about leisure time would be more meaningful.

To sum up this discussion of leisure time and its distribution in human societies, one can conclude that, defined as the sphere of life left over from the more necessary or fundamental social tasks, it exists everywhere, but is interculturally and intraculturally variable. There seems to be no question that the generation and allocation of leisure time are related to the system of production, but much more research needs to be done in order to make assured cross-cultural generalizations about these matters.

Leisure activities take many forms, one of which, tourism, involves travel. The *Outline of Cultural Materials* also considers travel a cultural universal. People in all societies venture away from home, even if only for a few hours, and as with leisure, the rate and nature of this travel is both cross-culturally and intraculturally variable. Tourism reflects the rate and kind of traveling carried out by the people of a society, as well as their leisure.

The social conditions associated with the tendency to venture abroad are even less well understood than those associated with leisure time. It has been suggested that traveling is linked with the complexity or development of a society; and with a little mental tinkering at the hunting and gathering end of the continuum, the argument that more developed peoples show a greater tendency to leave home looks inviting (though one also has to be concerned with intracultural variability here). Many authors have pointed to this tendency. Lerner (1958:247), for example, saw in the Middle East a general line of social development that had been taken in the Western World in which greater physical mobility resulted in "people becoming intimate with the idea of change through direct experience". However, the nature of the relationship between social complexity or development and traveling can hardly be spoken of with confidence at this point.

To tie together the preceding discussion, it has been argued that tourism involves leisure and travel, both of which have been viewed as socially generated. The social generation of leisure derives from a society's system of production in which people labor to produce the resources necessary to sustain themselves in their work and other activities. Whatever leisure time is generated by a society is allocated in ways that are consistent with its production system. The people of a society make use of the leisure time available to them in socially structured ways. One such way involves traveling. Whether people travel in their leisure time, and so become tourists, would seem to reflect (variably, of course) the general propensity for traveling in a society, which, itself, may be related to other socio-cultural factors.

So far in this chapter, only the minimal conditions or opportunities for tourism have been established. Whether these opportunities are realized, and in what form, would seem to be dependent on more specific factors to be linked with the variability of tourism within and between societies. Graburn (1983a:19–23) has proposed a list of factors "which generate the patterns found and which may predict touristic patterns". Though this list certainly is not exhaustive, and though its cross-cultural applicability remains to be determined, the factors which he has suggested can be used in hypotheses for further testing. For example, according to Graburn, cultural self-confidence is supposed to be associated with more or less adventurous forms of tourism. Studies in various cultures, using Cohen's (1979, 1979b) well-known classification of tourist types, which was formulated in terms of the experiential dimension of adventure-someness, can be conducted after the necessary adjustments for obtaining cross-cultural equivalences have been made. The testing of other hypotheses suggested by Graburn and others could follow.

The Notion of Superstructure

The view here that tourism is a kind of superstructure has been suggested by Sahlins' (1972:65) contention that leisure is "a superstructural counterpart of a dynamic proper to the economy". The habit of looking at the activities of a society in terms of base and superstructure is most closely linked with Marxist materialist analysis. Godelier (1980:5–6), taking a broad materialist point of view, see's the social base (or infrastructure) of a society, which consists of all of that society's productive arrangements, as exercising a determinative role "in the last instance" on all of its other activities. Harris (1992:297), also taking a broad view, argues that all features of a society are "necessary components of social life", but they do not "play a symmetrical role in influencing the retention of socio-cultural innovations". For him, the most influential aspects of social life consist of demographic, technological, economic and environmental features. For these two authors, it is the (somewhat differently conceived) 'material' base that has the ultimate influence in this regard. Other authors have argued that this or that aspect of society has a primary influence in shaping other elements of its social life at particular points in its history.

Though tourism may, in the last instance, be dependent on the materialist base of society, attempts to explain its variability, which have begun to preoccupy anthropologists, would seem to require more proximate, and therefore more varied causal hypotheses. It should not be forgotten though, that even though it ultimately may be a dependent phenomenon, like all aspects of a sociocultural system, tourism itself is capable of exercising some influence on other social activities, as, when for example the 'travel industry' exerts political pressure in favor of one of its interests or people who have visited a certain country begin to act politically on issues involving that country.

Nor is tourism to be conceived as some kind of mechanistic outcome of base arrangements. Rather, as Hardy (1990:542–543) argues for working class vacation sites in England, the individual (tourist) users "contribute to a greater extent than is commonly acknowledged, to the formation of distinctive patterns of use and opportunity" (in this regard, one can peruse with profit, Rojek's (1989) discussion of the issue of social constraints on the experience and action of individuals in leisure).

What processes connect the various social bases and touristic super-structures? In leisure studies two general connections between routine life at home and tourist behavior have been suggested (see Neulinger 1974:12). One has been called the 'spillover effect', in which leisure phenomena are a simple carry-over from base activities. Thus, alienated workers tend to seek various alienated forms of leisure, individualists are more likely to travel alone or in small groups and people raised in a consumer society contribute to the success of duty-free shops and other tourist shopping meccas. One of the better known proponents of this point of view in tourism studies is Boorstin (1964), who sees the super-ficialities of modern life spilling over into touristic activities.

The other proposed process connecting routine home life with tourism is compensatory. Thus, instead of continuing on the same home track, the alienated worker seeks a less alienated, more authentic existence during a vacation abroad, or the harried doctor looks for a place to unwind. In tourism studies, this point of view has become well-known through the work of MacCannell (1976) who sees the modern middle classes scouring the world for the authentic experience which has been denied them in an industrial–commercial world at home. It is easy for someone who accepts the existence of a universal need for alternation of experience – as does Graburn, whose position was critically examined in *chapter 3* (see also Nash 1984) – to subscribe to this point of view. Whether tourism involves a spillover from, compensation for a home routine, both, or something else should not be the main issue here, however. What is more important is the recognition that the causal connection between tourism as superstructure and its social base should be stated in a way that conforms to some notion of how human beings operate.

Whatever effect is at work in a given case is a product of a number of factors in the tourist generating situation. For example, people who are harder pressed at home may have an increased need for relaxation on their vacation; and as Cohen (1988b:376) points out in a critical elaboration of MacCannel's analysis of one of the inclinations of contemporary Western tourists, these people may seek more or less authenticity in their tourism according to how alienated they are from the social conditions in which they live. As far as leisure, and more specifically, tourism are concerned, Giddens' (1983, 1995) discussions of the ways in which social influences act on individual actors have had some influence on work in these fields.

Statements of the relationship between base and superstructure also should conform to some notion of human motivation, which, according to Pearce (1993:116), "is the global integrating network of biological and cultural forces which gives value and direction to travel choices, behavior and experience". A variety of motivational schemes for tourists have been proposed, with those offered by Dann (1981) and Iso-Ahola (1982) being, perhaps the best introduction to the subject. Dann argues that the general disposition to tour involves a push, while more specific dispositions such as destination choice involve a pulling component. Iso-Ahola, in a similar vein, speaks of avoidance and approach in his analysis of tourist motivation and sees tourist motivation in different societies as governed by more of one or the other.

Dann's analysis of what he takes to be an ideal–typical case is instructive. According to him, the logical and temporal sequence begins with desires which are not being satisfied at home, then a casting about for a way to satisfy them and a settling on travel as a way to do it. Destination choices, which constitute various 'pulling' alternatives, finally are narrowed down in some fashion, perhaps as is suggested by Crompton (1992) and Mansfield (1992). As Dashefsky and his colleagues (1992) have pointed out for overseas Americans, the picture in any specific case may be complex, with various factors weighing in from a number of directions, but the idea that tourists or prospective tourists are pushed and pulled in making their choices about touring, does seem to provide a beginning in understanding how social bases act through human motivational processes to produce particular tourism superstructures.

Some Tourism Superstructures

It may not be too early to offer examples of tourism studies in anthropology that follow the tentative, loosely-conceived base-superstructure paradigm that has been presented here. In the first place, some guidance is needed to help researchers respond to the calls, mentioned earlier, for anthropologists to attend to tourist generating situations. Secondly, the research that has been conducted on the questions raised in this chapter, though it may not have employed the base-superstructure paradigm explicitly, fits readily into this conceptual scheme. Two examples of research that does so will be examined here.

One study was conducted by Nelson Graburn, whose name has appeared often in this book. A pioneer in tourism studies in anthropology, he also has a long-term interest in Japanese culture. The study that will be reviewed reflects this interest. The other deals with American Indians – more specifically, how they have been depicted on picture postcards and in other literature. This kind of research, which Patricia Albers and William James have made their own, has excited scholars from a number of fields.

Graburn on the Social Organization of Japanese Tourism

As should be clear by now, Nelson Graburn has done much to set the agenda for tourism studies in anthropology. He is a broadly-trained sociocultural anthropologist whose interest in Japanese culture undoubtedly has been stimulated by his Japanese-American wife. His other specialties include the history of anthropology, Eskimos (Inuit), Indians of Canada and the arts, as well as tourism. As a teacher and administrator at one of the great California universities, he has influenced many students. His course on tourism is almost as well-known as his writings on the subject. Graburn has dealt in depth with other aspects of Japanese tourism, but it is mainly his treatment of its social organizational side in his first paper on the subject, *To Pray, Pay and Play: The Cultural Structure of Japanese Domestic Tourism* (1983b), that will be considered here. That treatment, as will become evident, is consistent with the notion of tourism as a kind of superstructure.

At the outset of his paper, which was published in one of the series offered by the important academic and reference Centre des Hautes Études Touristiques in Aix-en-Provence, Graburn (1983b:1) argues that there is a "necessary relationship between the two 'spheres of life' – work and home vs. leisure and travel". He treats the social organizational side of Japanese (especially domestic) tourism, its target sites (natural, cultural, religious, etc.), its timing (according to the yearly cycle of holidays and festivals) and style, as well as its historic basis in pilgrimage. Graburn (1983b:36–37) notes that the tendency of Japanese tourists to travel in large groups provides an obvious contrast with contemporary Western tourism. He thinks that this kind of touring – in large groups, rather than in small groups or as individuals – reflects the more collectively oriented nature of Japanese society.

It is perhaps because the more collective orientation of the Japanese is so well-known that Graburn does not spend much time demonstrating it. We do learn, however, about the various group involvements of the Japanese people. Kinship groupings, school groups, work groups and friendship groups play important parts in an individual's daily life; and in an interesting illustration of the spillover effect, mentioned earlier, they appear to influence tourists to act in a group-oriented way. Graburn mentions family trips to the seaside, amusement parks, etc., school tours, recreational outings of friends and tours sponsored by different work organizations (curiously, he refers only in passing to the tours arranged by travel agencies such as the huge Japan Travel Bureau). In all of these tour groupings, a 'team spirit' is considered desirable. To clinch his argument about the importance of travel in large groups for the Japanese, Graburn (1983b:39) says, "I have seen individuals rush to join a large group going somewhere, even if the person is not supposed to be going in that direction or to that destination".

There are other indications of the importance of the group in Japanese tourism. One is the relationship between travelers and their kinship groups at home, which is cemented with a variety of gifts. At a farewell party prior to departure, the tourist is given money, travel paraphernalia, or good luck charms (*senbetsu*), the value of which has to be carefully noted. During the trip, the traveler is supposed to buy appropriate return gifts (*omiyage*) worth approximately half the value of each *senbetsu*. These are delivered to the people at home at a welcome home party. The tourist also brings back souvenirs (*kinen*), including photographs, from the tour to the people at home to confirm visits on the tour. Further, the home group may carry out certain observances to keep the traveler in mind. As Graburn (1983b:46) points out, "...the traveler is 'sent' as a representative of an enduring group, most of whom are left behind, and indeed travels 'for them' and buys things that they would have bought if they had gone too". Thinking comparatively about those Americans, for whom travel is primarily a means of gaining or keeping status for oneself, may help to put this Japanese tendency into more profound relief.

Is Japanese tourism becoming more or less collectivized? The world is becoming more capitalized and industrialized, which suggests that, to the extent that they are committed to that world, the Japanese will be taking on a more individualistic way of doing things. When Graburn first wrote about Japanese tourism in 1983, increasing numbers of Japanese tourists had been venturing abroad. Japanese travel brochures aimed at this growing international market suggested to Moeran (1983) that the core clientele being appealed to for this kind of touring were younger, more individualistic and adventurous, upscale types. These would be the new wave of Japanese domestic and international tourism. He (Moeran 1983:95) then thought that "... the era of the flag bearing guide leading a party of bewildered Japanese tourists... [was] coming to a close".

Since the 1970s, there does, indeed, appear to have been an increase in touring by individuals and small groups of Japanese such as the family on both domestic and international fronts. This more individuated touring has been promoted by government policies "to respond to individual needs and tastes" (JNTO 1992:20), as well as the possibility of touring by private motor car on improved roads in Japan.

From the perspective of tourism as superstructure, this less collectivized form of touring can be seen as a reflection of an increasingly individualistic way of doing things on what Graburn refers to as the "serious" side of Japanese life. Hofstede (1980), in his study of employee-related attitudes in a large multi-national corporation, has shown that there is a tendency for employees from more complex industrial countries to be more individualistic; and all the evidence for industrializing Japan indicates that the Japanese are, indeed, becoming more individuated. A particularly telling example of this is provided by Schooler (1972:316–317) who found that though the Japanese were less individuated than Americans, already in the 1950s and 1960s, they were moving steadily towards a more individuated way of life; and in an interesting corollary to the later work of Moeran on tourism, he found this trend to be spearheaded by younger, higher-status Japanese.

The Japanese custom of touring in large groups, however, continues to flourish. Japanese airlines have added significant large-scale charter business, and Japanese travel bureaus continue to book significant numbers of tourists on large group tours. In one interesting example, Ehrentraut (1993:275) points out that "at least 30% of the visitors to heritage farmhouses in all accessible conservation categories are members of some kind of tour group of 20 or more persons". Then there are the notorious 'sex tours' of large groups of Japanese men, especially to southeast Asia, which, despite increasing criticism from many quarters, continue to flourish. The Japanese are not alone in this kind of touring, but somehow or other (possibly because of their collective presence), they stand out from others in this regard. In discussing them, Mackie (1988:223–224) says that "Japanese tourists can travel to the Philippines by Japan Airlines on a tour arranged by a Japanese travel bureau and stay in a Japanese-owned hotel".

The exact location of Japanese tourism on the individual–collective dimension is not clear. Neither is the exact nature of the Japanese historical trend towards individuation. Speculating about the future of Japanese tourism, Graburn (1983b:66) said that, even though there was a trend towards more individualized traveling, "at no point in history will the range of tourist structures and motivations coincide between Japan and the West". In a personal communication, he has reaffirmed this position, pointing out that the feeling of security provided by large groupings of Japanese travelers continues to be necessary for the support of a people who remain timid about leaving their home turf.

From the perspective of tourism as superstructure, the continuation of the Japanese collective orientation in tourism can be explained in terms of the basal structure of Japanese society, which continues to promote a collective orientation among these people. What is the relevant basal structure that explains the way in which Japanese touring is organized? Perhaps the collective orientation of Japanese in their daily lives at home, which Graburn relates to their touristic style, has to do with a more dense population; or possibly it is related to an earlier stage of industrial–capitalist development. Competing hypotheses certainly can be tested by interested researchers.

All of this discussion can be related not only to the debate on the trajectory of Japanese development in particular, but also the more general issue of development in the contemporary world (see McCormack and Sugimoto 1988). It can contribute to our understanding of sociocultural change by viewing tourism development in home societies as reflections of changes in more essential or basal aspects of a society. A host of other questions also arise when tourism is viewed in terms of the base-superstructure paradigm. For example, gender, class and ethnic differences in tourism, which have only begun to be investigated, can be seen as an effect of (basal) social positions of people at home, that in turn, act back to affect those positions in one way or other. Needless to say, market segmentation research (see Mazanec 1995) can be of assistance to anthropologists here in this kind of analysis.

Picture Postcards of American Indians

In a series of studies involving picture postcard representations of American Indians, Patricia Albers and William James have become well-known to scholars in a number of disciplines. Trained in anthropology, these anthropologists have conducted fieldwork in Latin America (James) and with various North American Indian groups (Albers). Their work on the photographic representation of ethnicity is based on a collection of more than 40,000 postcards from all over the world which have been accumulated over the years. Most of this work concerns ethnic groups in the North American west, particularly Indians, which has been continued by Albers since James' death.

The guiding view in much of this research is that tourism influences the way in which host groups are represented. In the discussion that follows, the focus will be on picture postcard representations of Indians of the southwestern United States around the beginning of the twentieth century. These representations, according to Albers and James, constituted an ideology that responded to the interests of White American society (see Albers and James 1983, 1988). Though they do not follow through this notion to the end, their work can be used to provide an illustration of the application of the notion of tourism as superstructure. Here, the author would like to finish their analysis off in a theoretical direction that seems to be latent in their work.

Albers and James (1983:131) argue that early picture postcard representations of Indians were mostly for local consumption and "showed a great deal about the ordinary and indigenous aspects of life...". With the growth of tourism, such images changed to meet the expectations of a wider range of visitors. In the southwestern United States, this change was especially apparent in the media productions of the Fred Harvey Company and the Santa Fe Railroad, which had virtual monopolies on the tourist business to that area. In these productions, the rather grim realities of daily life under White domination gave way to idyllic, exotic scenes apparently designed to meet the expectations of mostly White American tourists.

In a recent paper, Albers (1992:9) says that "the essence of the written and photographic language which accompanies the region's tourism has remained remarkable stable and highly romanticized". The enduring commonalities in linguistic and photographic discourse are viewed by her as media responses to perceived common denominators of an expanding tourist clientele which has expected the novel or exotic to be presented in a synthetic, sanitized way.

How, exactly, were such expectations generated in White American society? Albers and James really do not address this issue, which is to say that they do not explore connections between what have been referred to here as base and superstructure; but such connections can be suggested by referring to the image of the wilderness and those who inhabited it that was held by the American public. From the earliest days of the American Republic, the American Indian was a significant 'other' for

White Americans. At first threatening, the image of this 'other' became increasingly benign with time until, with the disappearance of the frontier and pacification of Indians towards the end of the nineteenth century, it developed into a romantic picture of the uncivilized.

This view appears to have been more prevalent among city-dwellers, who according to Roderick Nash (1967:231), had the greatest reason to question "the beneficence of civilization's achievements". Perhaps Krippendorf's (1986:523) comments about contemporary peoples' urban life will give readers an idea of what these Americans had begun to experience. He says:

> Their work is more and more mechanized, bureaucratized, and determined without regard to their wishes. 'Deep inside', they feel the monotony of the ordinary, the cold rationality of factories, offices, apartment buildings, the highway infra-structure, the impoverishment of human contact, the repression of feelings, the degradation of nature and the loss of nature.

That the images of the American Indian held by the American public did, indeed, reflect an increasingly popular enthusiasm for the primitive is suggested by the ideology of the Boy Scouts of America, already the largest youth group in the country by the turn of the twentieth century. Its Handbook for 1910, written by E. T. Seton (1910:1–2), emphasizes the value of living "the simple life of primitive times". Theodore Roosevelt, who seems to have been a kind of boy scout all his life (see Miller 1992), put the power of the United States' presidency behind such views. Thus, the pristine images of American Indians found by Albers and James in their later postcards seem to be in line with an increasing enthusiasm for a world that contrasted with that being produced in industrializing and urbanizing America, a world conceived in a romanticized way.

What was the function of this romantic ideology for a United States that was becoming ever more industrial and commercial in its way of doing things? A number of authors have argued that images of a beneficent state of nature provided some relief from the human problems found in such a society by offering an opportunity for social rejuvenation. Heiman (1989), for example, speaks of the social contradictions in the American capitalist base in his study of conceptions of the Hudson River Valley, the most famous views of which are to be found in the paintings of the Hudson River School. The contemplation of a highly romanticized nature is seen by him as a way for American society to, in effect, heal itself. Postcard and other images of American Indians also may be seen as a rejuvenating response to the increasingly oppressive business or industrial side of American life.

One may pursue the analysis opened up by Albers' and James' studies of picture postcard representations of American Indians still further by arguing that the romantic American ideology, just discussed, was variously interpreted by tourists and potential tourists, each having their own motivations (ego enhancement, relaxation and the like). Such motivations would constitute 'push' factors that, in this case, responded to the 'pull' of pacified Indian territory that, according to Albers and James (1988:151–152), was given a "picturesque, exotic, and enchanted character" in the media productions of agencies such as the Fred Harvey Company and the Santa Fe Railroad.

Finally, whether the psychological dispositions engendered by this American ideology were translated into a decision to travel and a subsequent choice of Indian country or something like it as a destination would involve some kind of process of narrowing down such as that suggested by Crompton and Mansfield, which, as a number of authors have pointed out (see e.g., Boyer 1972:45–58; Jones *et al.* 1983; Haukeland 1990), is shaped by the situational constraints of daily life at home.

Thus, in this little overview based on the picture postcard research of Albers and James a causal connection has been suggested between base and touristic superstructure of a developing American society. It should be evident that this proposed connection has been seen as a kind of compensation for social problems in that society. In addition, the facts of individual motivation and the relevant social context have been used to suggest how all of this is realized in the choices of individual tourists of a particular tourist destination. One could carry this analysis still further by suggesting the influence of all of this on the tourist region of the American southwest, which according to Weigl (1989), has become a kind of giant Disneyworld that was originally developed to meet the interests of touristic agencies like the Fred Harvey Company and the Santa Fe Railroad and their clients. Nash (1979b:68) states the principle involved here when, in connection with his analysis of the development of a tourist culture in Nice, he says that "an understanding of the touristic metropole and the tourists it generates is particularly important for comprehending developments in a tourist satellite or resort".

The two summary case studies and discussion just presented may have raised more questions than have been answered. By way of exculpation, it should be pointed out that this presentation has been designed to open up a new line of research, and so is exploratory. Later studies seeking answers for questions about other qualities of tourism, such as, for example authenticity or commoditization, will have to employ more adequate research procedures to answer more specific theoretical questions. Though it is the custom for anthropologists to speak about whole societies, it is obvious that one has to be particularly careful about doing this in big, highly differentiated societies such as Japan and the United States. Anthropologists having an interest in the generation of tourists and tourisms might be well advised to concern themselves initially with smaller, more manageable sectors of a society while still being guided by the base-superstructure paradigm. Data used in such a research program will have to be much more adequate, and anthropological strictures that encourage seeing things in their social context, understanding the point of view of subjects and keeping alive the possibility of cross-cultural comparison should be kept in mind.

In this regard, Cohen (1993), who has been in the forefront of tourism research from the beginning, has been working on an interesting study of the generation of various kinds of images of native peoples (art work, as well as photographs) for touristic consumption. His work raises questions about who is producing the images, for whom, how, in what context, in what medium, etc. The different images of the Indians of the southwestern United States can be analyzed in terms of his cross-culturally applicable scheme. For example, among the approximately sixty different agencies that have produced photographic representations of the Navajo Indians, some are owned by the Navajo themselves. How do the images produced by Navajo-owned agencies compare with those produced by Whites, and why? Cohen's system of analysis would help us to answer such a question.

Both of the summary case studies used in this chapter have been offered to explore the feasibility of carrying out anthropological investigations of tourism at the tourist generating end of the touristic process. From the perspective of tourism as superstructure, they provide tentative causal explanations of different forms of tourism. The explanations that have been sought here concern, not leisure, travel, or tourism in general, which were considered at the beginning of the chapter, but rather, different forms of tourism. Questions have been asked about why the Japanese have tended to favor a more collectivistic form of tourism and why the Indians of the southwestern United States were represented on picture postcards in a certain way. A different basal factor was seen to be operative in producing some aspect of tourism in each case. No attempt has been made so far to determine whether the basal factors suggested provide the best explanation for the aspects of tourism being considered.

In the Japanese example, the proposed social base for the collective orientation of Japanese tourists is the way in which Japanese society is organized (other, 'deeper' factors such as the great density of Japanese living arrangements or the state of development of Japanese industrial capitalism have also been suggested). That way has been more collectivized than in the United States, but it appears to be becoming more individually oriented. The mostly impressionistic data available suggest that the social organization of Japanese tourism reflects this basal aspect of society. Further research should be able to get at the specific basal factor that offers the best explanation of the phenomenon.

In the Japanese case, the connection between base and touristic superstructure was seen to operate as a kind of spillover effect. This is not to say that all Japanese tourism can be viewed only as a spillover from some basal aspect of ordinary Japanese life. Indeed, Graburn (1990, 1995) himself, has argued that Japanese tourism to rural areas such as spas and heritage sites results from an attempt to find some kind of nostalgic rejuvenation in the face of the increasing problems of living in a modern, urbanized society. This suggests that, unlike the organizational side, this aspect of the Japanese touristic superstructure represents a compensation for contradictions in the basal routine and leads to the conclusion that even as a society's touristic superstructure is variable, so too are its basal causes and the processes that produce them.

In the picture postcard example, the base-superstructure connection is somewhat more complicated. Tourist agencies are involved in the marketing of a destination area in a process that has been studied by Silver (1993) and Riemer (1990). The representation of Indians on the postcards has been seen to be a creative response by the Fred Harvey Company and the Santa Fe Railroad to their perception of the views of Indians by their potential or actual tourist clientele. Using one possible line of explanation, it was suggested that those views reflected an interest in the exotic which stemmed from a desire to get away from an increasingly oppressive industrial and commercial society. The social production of such views, therefore, is an example of the compensatory effect of a tourism which 'makes up' for some inadequacy in ordinary social life. It may be, however, that the 'style' in which the Indians are represented also is a spillover from the way in which things are pictured at home, that is, in an increasingly 'commoditized' way that embraces ever broader common denominators.

* * *

The perspective of tourism as superstructure provides only a general orientation for explaining the generation of tourists and tourism. Within this general perspective, a variety of theories about how a society works may be tried on. It is important to keep in mind that the theoretical approaches suggested by the author in the summary case studies presented in this chapter are not the last or only word about the basal factors involved in the generation of these specific forms of tourism. To cite some of the various other ways that might be used to explain the same phenomena, one can take touristic images as kinds of myths, as Barthes (1972:74–77), in his analysis of the *Guide Bleu* and Dufour (1977), in his analysis of French mythic views of the weekend, have done; or one might develop some variation of the theory of basic personality type (Kardiner 1939, 1945), which would look at tourism as an 'expression' or 'projection' of a society's basic personality; further, one could consider the struggles of social classes, as Bourdieu (1984) has done, and see different forms of tourism chosen by the different classes as reflections of these struggles; and finally, one might attempt to extend any one of the theoretical approaches to leisure analyzed by Rojek (1985), formalist, neo-Marxist, deconstructionist and figurational in pursuing an analysis. Of course, whatever theoretical approach is employed must be backed up by more adequate methodological procedures than have so far been demonstrated in most anthropological studies of tourism.

To end this chapter on a more down-to-earth note, consider an early study by Valene Smith (1979), employing typical anthropological procedures, of the motives involved in the choice of tourist destinations. Among the data gathered are those which refer to what may be construed as the social base of these choices. Smith's study, which unaccountably has not been followed up, made use of information routinely gathered by a California tourist agency in working out client itineraries, as well as the anthropologist's observations of tour planning sessions and her interviews with the clients. Showing her anthropological nature, Smith (1979:54) reports that she was "wary of the questionnaire approach in trying to get at the reasons for touring decisions".

Most anthropologists are likely to feel more at home in small-scale settings where the possibility of give and take with subjects is greater. This seems to have been the case for Smith in working with the itineraries of 141 travelers from a small town and in a travel agency with which she was intimately connected. Her research procedures offered her the opportunity to empathize with her subjects. In keeping, perhaps, with another aspect of her anthropological nature, the study is only lightly theoretical. One can, however, see possibilities now for phrasing her interesting account in terms of the notion of tourism as superstructure.

Smith provides tantalizing references to what could be the basal aspects of the tours chosen by her subjects. Some psychologically-oriented anthropologists might see the process of working out an itinerary with a travel agent as a small part of the socialization or enculturation process that begins in childhood and is shaped by an ongoing series of social contexts. Smith, herself, suggests the social contexts that different tourists had in common at the time they were making their choices, thus pointing to more or less salient basal aspects of the tourism being practiced in her community. For example, she indicates what amounts to a collectivity orientation among peers (not, it should be mentioned, so strong as that of the Japanese), which appears to produce agreement about where to go, helps to explain the emphasis on Hawaii among them and could also account for subsequent shifts among these people to other 'in' tourist destinations.

The well-used (in anthropology) perspective of socialization or enculturation, that traces the development of the human individual is only one of a number of possible approaches for seeking out the basal elements on which particular kinds of tourism are grounded. It does, however, appear to be particularly congenial for many anthropologists, for whom the project of getting to know the 'other' and explaining his or her actions goes best in a smaller-scale setting. Anthropologists interested in research on tourism as a kind of superstructure certainly ought to find projects that follow from Smith's initiative rewarding.

Chapter 5

The Anthropological Approach to Tourism: a Preliminary Assessment

In a pathbreaking article, A. Irving Hallowell (1955:89) pointed out that "the human individual must be provided with certain basic orientations in order to act intelligibly in the world he apprehends". At home people are normally given a fairly stable orientation in their culture. They don't have too many problems figuring out where they stand. On a trip, that stable, taken-for-granted orientation can be disrupted so that they begin to question themselves and their relation to the world. Then, they may feel the need to stop and reflect about these things in order to better cope with the uncertainties of an increasingly dislocated existence. Perhaps they will be aided in doing this with representations of the progress of their trip on a video screen or discussions of the peoples and places en route. An airline pilot, tour guide, or experienced fellow travelers also can help them to adapt to the dislocations brought about by touring.

As far as this book is concerned, this is a time for making better sense of where we stand in our tour through the field of anthropologically-oriented tourism studies. What has been the course of this tour of the field so far? As is to be expected, individual readers will have different ideas about what has been covered. Each person will have begun to identify what are, for him or her, more significant points of reference and to form tentative general impressions. The author, who is a kind of guide in this tour, has his own notions, which are, perhaps, more privileged because he has planned the trip and knows, in a general way, how it is going to come out. The first part of this chapter will be his summary of the main points in what has transpired so far.

Chapter 1 was an introduction to anthropologically oriented research on tourism. That research, as subsequent chapters have shown, seems finally to have taken off. Looking back from a vantage point several decades on, it appears that anthropologists have contributed a good deal to our understanding of tourism. As in the history of any culture, the growth of this field of study shows false starts, but also lines of development that appear to be fruitful for future study. Aided by a critical analysis of what has transpired, it should now be possible to sort out promising procedures and paths for future researchers.

What we know now about tourism from an anthropological point of view has been arrived at through distinctive anthropological procedures for working with and understanding people, comparing the life ways (cultures) of groups of human beings with each other and putting everything in a context that can include all of humankind. Like other scientific disciplines, anthropology has both a basic and applied side. Basic science has no extrinsic, practical agenda. Research is not undertaken to help solve this or that human problem, as in applied science, but rather to solve cognitive problems generated by scientific attempts to comprehend the world.

For anthropology, the issue of trying to work with, and understand the numerous and varied human 'others' who constitute its field of interest, is currently a bone of contention. The study of people poses special procedural and ethical problems. Anthropologists are also committed not just to trying to describe and analyze human behavior, as the behaviorist sees it, but also the meanings which human actors attach to it. What, for example, are the poor among the Burmese up to when they donate what often appears to be an excessive amount of their incomes toward the building of pagodas? What did the Trobrianders intend in engaging in the exchange of non-useful valuables in the famous *kula* trade? This problem of understanding the 'other' is sufficiently great to cause some anthropologists to seriously question the scientific status of their discipline.

Many anthropologists, however, continue to have faith in their ability to understand those 'others' they are studying (see Spiro 1992 for an elaboration of this viewpoint) and make generalizations about their actions. In tourism study, this means the tourists, all those who serve and deal with them and others on the fringe of their activities. In the ideal–typical sense, basic research in the anthropological study of tourism is aimed at comprehending this field of human action, nothing more. The first half of the book has looked into anthropological attempts to understand various kinds of touristic actions in different culture contexts.

The applied side of anthropological research on tourism, that is, the side which attends to practical questions such as the sustainability of a given course of tourism development or the value of a particular orientation program for tourists, will be considered in chapters to come. That applied side has contributed significantly to the anthropological study of tourism. Indeed, the famous seminar sponsored by the World Bank and UNESCO (de Kadt 1979) can stand for what seems to have been the predominant thrust of anthropological work on tourism, that is, the attempt to discover ways in which tourism can be made to benefit hosts in the less developed world. Indeed, much of the research that anthropologists have conducted from the perspective of acculturation or development, has been carried out with the explicit or implicit agenda of helping host peoples who are experiencing the impact of tourism on their societies and environments. This research has helped anthropologists too. One should not overlook, for example, that it has paid the way for some of the anthropologically oriented work on tourism.

Applied research on tourism certainly can contribute to our comprehension of this subject, but the introduction of practical considerations puts a special spin on research which makes it advisable to consider it apart. Such a consideration will be offered in *chapters 6* and *7*. After that (in *chapter 8*), it will be time to look into the question whether the two sides of the anthropological study of tourism – the basic and the applied – can move forward hand in hand, and if so, how.

For the moment, consider the basic side, which we have been exploring in the first part of this book. From the work of anthropologists and other social scientists, we certainly know more about tourism now than we knew when Nuñez (1963) introduced the subject in anthropology; and thanks to the efforts of some anthropologically-oriented scholars, we know about it not only as a contemporary Western phenomenon, but in other cultures as well. Not only do we know more about tourists and their practices, we also know more about the consequences of these practices for hosts and the tourists themselves. Anthropologists also have made some small beginnings in coming to terms with the processes involved in the social generation of tourists and tourisms, which is to say, its causes. Much of all of this work can provide a foundation for future research.

Like any developing field, anthropological study of tourism has had its growing pains. As was pointed out in *chapter 2*, early research, which largely concerned the consequences of tourism development for societies in the less developed world, contained a good many (often negative) value judgments. This negative view of tourism's potential for aiding the well-being of host societies (not always shared by host peoples) may have sprung from a mistrust by many anthropologists of the establishments of their own (Western) societies and the imperialistic activities of these societies (and perhaps, their discipline) in the less developed world.

True, a few anthropologists were advocates of tourism development, and some would become advocates only after being assured that adequate planning and controls in the interest of hosts were in place; but too often tourism was (and still is) considered as something good or bad, not as a subject to be studied simply to be understood in some scientific way. If one already 'knows' the value of something one is investigating, there may be a tendency to slough off diligent scientific work. Anthropological and scientific strictures may lose some of their force. As Greenwood (1989:183) says, in his own review of his earlier work on the tourism-induced changes in the Basque *alarde* festival of Fuenterrabia, "Moral anguish was easier to express".

Now, it does seem that the anthropological regard for tourism has, indeed, moved beyond the stage of reflex-like moral responses to what Jafari (1990) has called the "knowledge based platform" of tourism study, but even so, one still cannot be easily sanguine about scientific progress in the field. Descriptions have proliferated and theories have been offered, but according to Dann *et al.* (1988:4), "there has been an unfortunate tendency to gloss over questions of theory and method and a concomitant failure to acknowledge their interrelationship". Based on a review of work in two of the leading journals in which social scientists publish their studies of tourism, these authors (1988:10) conclude that in recent years there has been a movement away from purely descriptive or theorizing articles on tourism, but that the "happy state" in which high theoretical awareness is combined with high methodological sophistica-tion has not been reached. The specific trend for the anthropological study of tourism does not appear to be out of line with this generalization for all of the social sciences.

The well-known work of MacCannell (see especially 1976, 1989 and 1992) provides an interesting illustration of issues raised by Dann *et al.* (1988). This American sociologist-anthropologist reports that on a visit to France in the early 1970s he was transformed from a rank empiricist into a theorizer in the French tradition. His subsequent theoretical adventures have led him in the direction of 'critical theory' or 'cultural studies', which have strong interpretivist leanings. In his writings on tourism, he has brought forward a number of useful concepts for the study of tourism in the contemporary world. For example, his name (see e.g., MacCannell 1976:91–108) often is associated with the concept of authenticity. For him, tourists in the Western world are pictured as looking for a sense of authenticity that is lacking in their life at home; further, abroad, there are hosts who are trying to provide it for them by creating authentic-seeming presentations of their own culture (staged authenticity). The develop-ment for tourist visitation of a Korean village of yesteryear outside of Seoul, South Korea, or the fabrication of 'traditional' crafts for tourists by Eskimos are examples.

This stimulating hypothesis has opened the door to a line of inquiry that has fulfilled some of its promise while revealing certain inadequacies. Cohen (1988b), for example, has pointed out that only some Western tourists pursue authenticity. There are, indeed, other kinds of tourists (and, one might add, hosts) for whom the issue of authenticity does not come up. He goes on to show the theoretical possibilities that emerge out of the recognition of this fact.

It would be a mistake to conclude, however, that because of empirical inadequacies, MacCannell's work is methodologically unsophisticated.

His French and subsequent experiences may have led to a less systematic concern for what is actually 'out there', but he now honors hard-headed ethnography when he (MacCannell 1992:308) says that we must have "the combination of 'critical' and 'observational' talents". His views on observation are sophisticated. He thinks that it is possible to understand the world, but probably not in an unbiased way (MacCannell 1992:1–13). Consequently, it is important for the researcher to take an effective position for observing subject matter and to declare as fully as possible where he or she is coming from. According to MacCannell (1989:3), any current social science "will render accurate accounts only to the extent that it keeps a clear eye on potential bias that might come from its own institutional, class, and historical position".

This view is consonant with the growing interest by anthropologists in the observer as one of the factors that contribute to the portrayal of 'others' (see Nash and Wintrob 1972, for an early analysis of this trend). Such a self-reflexive position, which now seems to be well established in anthropology, still has not made much of a mark in tourism studies, however, including those by anthropologists, where a somewhat naive empiricism continues to reign (but see Bruner 1991). In this regard, MacCannell, as a student of tourism, seems to be ahead of his colleagues in the field. His reflexivity does not appear to extend to research procedures, however.

Considering all this, where does the anthropological study of tourism stand as a science? In the formative years of this field of study, there was, indeed, a good deal of pure description and pure theorizing, often with a moral component. This continues, but there seems to have been some movement away from moral considerations towards a more objective understanding of tourism for its own sake. Descriptions have multiplied in work that is gradually becoming more theoretically informed and methodologically sophisticated, and the work of generalization has begun. One would be justified in expecting the anthropological study of tourism to continue to follow this path in the future – more or less depending on how much give and take there is with the mainstreams of anthropology and other social scientific disciplines, as well as the tourism industry.

The Look of Tourism: an Anthropological View

In *chapter 1*, some attention was given to the concepts of touristic process and touristic system, which were created for the analysis of tourism wherever and whenever it occurs. Those concepts are derived from the definition of the tourist as a leisured traveler, a conception which, as it turns out, leads to the identification of tourists in societies at all levels of social complexity. From this perspective, tourism is not myopically viewed as something that has evolved only recently in the West and spread to other cultures, as is often claimed, but rather a widespread human phenomenon that has varied cross-culturally and intraculturally throughout human history and pre-history. Using this conceptual base, it is possible for anthropologists to make the kind of cross-cultural comparisons (including, perhaps, all of the cultures of an evolving humankind) which are one of the hallmarks of their profession.

The scientific project for studying tourism is to describe it as objectively as possible in all of its manifestations, to order these descriptions in some kind of comparative way, and to identify its cause(s) and consequence(s). The anthropological project is not only to do this, but to try to comprehend tourism's many manifestations from an anthropological point of view, which means doing one's best to understand and analyze human actions in the contexts of touristic processes wherever they occur. Touristic actors may be conceived as playing various roles that include tourists, transport personnel, hosts and the like. In anthropological studies of tourism, one or another of these actors have occupied center stage, depending on the point of view that has framed the research.

Three general perspectives that have guided anthropological work in the field have been identified here. Anthropologists studying tourism have tended to approach tourism as a form of acculturation or development, as a personal transition for the tourist and as a kind of superstructure. These perspectives and what they have revealed were examined in *chapters 2, 3* and *4* and will only briefly be touched upon here.

Tourism as Acculturation or Development

The perspective from which the anthropological study of tourism got its start, that is, the acculturation or development point of view, emerged when anthropologists became aware that the societies they were studying were in the process of changing – often dramatically – as a result of contact with the Western world. As Nuñez (1989:265–266) points out in the revision of his seminal paper on the anthropological study of tourism, "Anthropologists have known for more than a half century many of the things likely to eventuate when different cultures come into contact, and this knowledge can readily apply to contact between tourists and indigenous or 'host' societies". Nuñez goes on to mention that the borrowings engendered by such contact (between representatives of more and less developed cultures) tend to be asymmetrical, the reason being that the more developed culture of Western tourists dominates, by one means or another, that of the lesser developed hosts. Following from this is the notion of tourism as a form of imperialism (see e.g., Crick 1989; Nash

1989), which has struck a responsive chord among those anthropologists who have a tendency to side with the underdogs of this world.

Only a little acquaintance with the subject of tourism will show the limitations of this image of tourist contact. Indeed the kind of cross-cultural contact posited by Nuñez, that is, the tourism between what has been variously referred to as the developed and developing world, Western and non-Western, center to periphery, industrial to pre-industrial, or north to south, seems to apply to only a small percentage of contemporary international tourism (see Høivik and Heiberg 1980:70).

Anthropologists have tended to see tourism transactions as involving significant power differences in favor of the more developed tourist-generating centers. No doubt, the power difference between cultures in contact is important, but to argue that it takes only this one form in all tourist–host contacts, as the author has done (Nash 1977, 1989), is to obscure a good deal of its cross-cultural variability. What anthropologists who are committed to a full-scale investigation of human life should be prepared to do is study all kinds of touristic contact in all of its social contexts. Certainly, an important dimension of such contact, wherever it occurs, is the power differential between the peoples involved, but it cannot be assumed that everywhere it conforms to the same imperialist model that has been prevalent in anthropological studies of acculturation or development. Moreover, there are other aspects of this contact that need to be investigated as well.

For pursuing such a project, the New Zealand geographer Douglas Pearce offers a framework for making the kind of comparisons that often are a part of anthropological work. In his scheme, it is possible to compare cases of tourism development in terms of a number of factors, one of which is the amount of local or host control or power over that development. On the basis of a quick overview, he is able to offer the jarring conclusion that "local control is not necessarily the 'good thing' that many writers imply, particularly where that control is in the hands of development-driven politicians" (Pearce 1992:26).

Inasmuch as the Pearce scheme recognizes what appear to be the more important factors governing tourism development (context, nature of the tourism, facilities, identity of the developers, etc.), it offers anthropologists an opportunity to begin to isolate the consequences of different kinds of touristic contact and so move forward in formulating statements of causation (for example, that certain kinds of tourism result in a widening of the gap between the rich and the poor among a host people) according to the perspective of tourism as acculturation or development.

Of course, as was pointed out in *chapter 2*, there are methodological problems in moving toward such causal statements – problems which anthropologists have been slow to recognize; but there is no reason to believe that, once recognized, these problems cannot be overcome. One is hopeful that the trend towards more effective experimental controls and more sophisticated statistical techniques that help to isolate the effects of different factors on tourism development will continue in the future.

Tourism as a Personal Transition

In looking over the advertising for tours to "Third World countries", the kind of tourism that anthropologists have tended to dwell on, Bruner (1991:239) finds that it offers the Western individual "nothing less than a total transformation of self". Natives, on the other hand, are depicted as being suspended in some "prehistoric" or "primitive" state, "untouched and unaffected by social changes". From what has been said about these matters in *chapters 2* and 3, it should be evident that these advertising claims are misleading. In regard to changes in the parties involved in touristic and other transactions between the Western and non-Western world, anthropologists have become wiser.

How wiser? Certainly, there now should be an understanding that, as a result of this kind of contact, the bulk of the changes should occur in the Third World peoples. Further, it should be evident that there are different kinds of touristic contacts in different contexts with different consequences for the people involved. Another thing that is coming to be taken for granted in anthropology is that, prior to the introduction of tourism into their societies, Third World peoples were not living in some pristine aboriginal state. Not only had they been in contact with other Westerners (governors, traders, missionaries, etc.), but also with each other. What we know of the history of the San peoples of southern Africa, who, in their hunting and gathering state, often have been taken as models for the most ancient humans, is important testimony to this. Those San, it appears, actually were pushed into what was taken to be an aboriginal condition in the Kalahari desert from a previous more generous environment.

MacCannell (1992:283–309), in his latest book, alludes to this issue in discussing tours of Westerners to places like the Sepik river in New Guinea, another region populated with peoples who often have been taken to be among the most primitive in the world. The tourists here have come mostly from Europe and represent what MacCannell refers to as a post-modern culture. The natives, whom many of these tourists take to be real primitives, actually are, according to MacCannell, "ex-primitives" who already have acquired post-modern ways (but not, perhaps, as wholeheartedly as MacCannell implies). So, the two sides meet in a kind of never-never-land – the 'empty meeting grounds' of his title.

It might be a mistake, however, to get carried away with MacCannell's scheme. He is, after all, talking about only a very small category of touristic contacts. Very few Westerners undertake this kind of touring. Those who do certainly encounter hosts who vary in their degree of acculturation. On the other hand, the fact that tourists, being tourists, are *en passant*, a condition that, itself, tends to lead to superficial social contacts, might justify the broader use of the term 'empty meeting grounds' for tourist–host contacts generally, not just in encounters between so-called ex-primitives and post-moderns.

On the tourist side, recalling *chapter 3*, one ought to be extremely cautious about accepting any claims for dramatic experiences or transformations of self, if for no other reason than the fact that the tourists on such tours, not only are *en passant*, but usually spend much (most?) of their time in the company of other Westerners and with access to Western amenities (as on boat tours to New Guinea). They have only comparatively brief contact with natives who may present themselves to tourists in the artificial way that MacCannell has referred to as "staged authenticity". Any dramatic transformations would seem to be confined to only a few of the most adventurous kind of tourists, perhaps those who Cohen (1979b:189–193) refers to as "existentials".

What the tourist gets out of his or her tour will depend not only on the kind of tourism being practiced, but also a number of factors associated with traveling and contact with hosts. Research on the consequences of different forms of touring on tourists, unfortunately, is not very far advanced, but the idea that tourists in general, driven by some undemonstrated need or other, will get carried away and be significantly changed or renewed by their experiences does not, on the basis of the review of this subject in *chapter 3*, appear to be justified even as a heuristic device.

Keeping this in mind, it would seem better to begin research on the subject of tourist experiences and reactions with the assumption that they will differ according to the nature of tourism they practice and a number of factors that figure in the context of their tour, some of the more important of which were discussed in *chapter 3*. These include whether the tourist is alone or traveling in a group of peers, the presence of a guide or guides, the nature of the host culture and the relation of the tourist to it (in which power differences are particularly significant), the nature of the tourist's home culture and the relation of him or her to it, and how many similar strangers have descended on the hosts and at what rate. Some of these contextual factors obviously overlap with the nature of the tourism being practiced. For example, mountaineering tourists in the Himalayas have a good deal to do with certain native guides, cooks and porters while those on a sedate 'cultural' tour of museums, architecture and music festivals will tend to have little direct contact with hosts because a guide (who may or may not be a native) often serves as an intermediary.

Suggestions for developing such a program of research on the consequences of touring for the tourist were included in *chapter 3*. What was not
considered there, except in passing, were the social consequences of the
tourist's touring on the society to which he or she returns. Do the
experiences and reactions associated with individual or multiple tours
spill over into these home societies? If tourists are recreated or renewed
even a little bit by their touring, as Graburn has suggested, what is the
effect of this on their home society? Did the Mather brothers and other
Calvinists from Boston, who visited the spa village of Stafford Springs,
Connecticut, indeed return from their 'vacations' there with renewed
energy for the daily routine? In other words, did their vacation actually
function as it was supposed to according to the ideology of their society?
Does a visit to Disneyworld, in fact, help to support the ideals of the
'American way', as Rojek (1993) has maintained? And what about the
probable return of those aging retirees who have spent years after retirement on the southern coasts of Spain? Jurdao and Sanchez (1990:161–
170) speculate about the return of many of these tourists to their home
countries in the north to end their days. What will be the social consequences of this return? The (in anthropology and sociology) well-used
concepts of social function and dysfunction (for the home society),
though under a good deal of critical pressure, are some of the conceptual
tools at hand for carrying out this kind of analysis.

Tourism as a Kind of Superstructure

In *chapter 4*, a way that has so far been little traveled in the anthropological study of tourism was explored. That way starts out by taking the
situation where tourists are produced (in economic terms, the demand
side as opposed to the host supply side in the destination area) as the
point of main concern. From this perspective, it is possible to begin to
comprehend the social generation of tourism, which is to say, its cause or
causes. Here, where the tourist or tourist-to-be has grown up and lived
most of his or her life, the major forces that create tourism or specific
kinds of tourism are to be found.

When Boyer (1972:19–58), in his pioneering work on tourism, used
social survey data for comparing rates and kinds of tourism undertaken
by the French, he for one, provided a basis for this kind of investigation.
Turning anthropologists' attention to the tourist generating situation
ought to be encouraged by the fact that it is becoming more and more
acceptable, and occasionally, even desirable for them to consider the
peoples of the Western world, which happens to be the place where they
and most of the tourists they have considered come from.

Taking tourism to be the activity of people who are at leisure and who occupy their leisure time with some kind of traveling, questions were asked in *chapter 4* about the causes of these two aspects of tourism. In this analysis, the productive system and traveling arrangements of a society were considered to be crucial. Though the exact nature of the causes of leisure time and traveling (hence, tourism) have yet to be discovered, it was argued that, given what appears to be the socially dependent nature of tourism on other social institutions 'in the home society' (it may be the opposite in some 'host' societies, as, for example the U.S. Virgin Islands, where 'received' tourism tends to dictate to other institutions), the well-used and suitably broadened concept of superstructure can be employed to study its generation. As a kind of superstructure, tourism may be seen to be dictated ultimately by the basal productive and traveling arrange-ments of a society, though in keeping with the more or less integrated nature of a society's way of life (that is, its culture), it also should be seen as capable of acting back on these arrangements.

From the accumulating work of anthropologists and other social scien-tists, it is obvious that tourism takes many forms. One can immediately appreciate this when considering all the different kinds and rates of traveling at leisure and the different kinds of activities that tourists undertake. What are the causes of the variability of these forms within and between societies? It is with the question "Why different rates or forms of tourism?", rather than the more intractable "Why tourism?", that anthropologists are likely to be concerned.

One way of getting an answer to such a question is to examine the touristic practices of different groups in a tourist generating society. For example, Urry (1990:45–46), an English sociologist, looks at what seem to be enduring differences between social classes in Western society. He notes an apparent tendency for upper-status Westerners to prefer soli-tude in their touring, perhaps of the kind analogous to what is depicted in the romantic landscapes of the American Hudson River School of painting, while the working classes tend to prefer a more collective mode with lots of people around, as on the beach at Blackpool in England. Is it, in fact, social class that is responsible for this apparent difference, and if so, what is there about the living arrangements of the different classes that brings about these different touristic inclinations? Information about the social and psychological conditions of the peoples involved, a good deal of which already exists, should help to further such a project.

Another way of answering the question "Why different forms and rates of tourism?" is to compare modes of tourism in different tourist generat-ing societies. In *chapter 4*, a summary case study was included about a line of research, begun by Graburn (1983a, 1983b) in which the collective orientation of Japanese tourists (a preference for traveling in large groups) was compared with the more individuated orientation of tourists from North America and Europe. On the basis of Graburn's analysis, these different modes were seen to be spillovers from what Graburn referred to as the more "serious" side of social life at home. As far as the Japanese are concerned, that more serious side is filled with a wealth of collective involvements in which the Japanese learn their collective orientation. In both the Urry analysis of social class differences in Western tourism and Graburn's investigation of a cross-cultural difference between Japanese and Western tourists, differences in modes of tourism were interpreted as being the result of basal sociocultural arrangements at home, the specific natures of which call out for further investigation by anthropologists.

The broadly conceived notion of tourism as a kind of superstructure provides a basic framework, within which it is possible to try on a number of theories from anthropology and other disciplines to attempt to answer specific questions about the generation of tourists and tourisms. Some indication of how this could be accomplished was given in the summary case studies of Japanese tourism and the picture postcard representations of North American Indians of the southwest at the end of *chapter 4* and the critical analyses that followed. As Graburn (1983b:55–66) points out in the conclusion to his study, this type of analysis can help us not only to explain or account for different forms of tourism, but also how a tourist's home society actually works. That society often turns out to be an anthropologist's own ought not to inhibit such an investigation.

Towards a Bigger Picture

So far, the anthropological study of tourism, despite delays in getting going – delays which seem to have been related to a general lack of respect for this subject in the culture in which anthropologists come from – has developed to a point where general trends are visible and substantial critical analysis can be undertaken. The study of tourism from the anthropological perspective has been carried out from three broad points of view which have provided the basic framework for anthropologists' approaches to this multifarious subject.

With this in mind, it may not be too early to suggest a broader conceptual scheme that will provide a context for these, and possibly other perspectives for viewing tourism and give anthropologists the broadest possible working view of the touristic field. This grander view can be provided by expanding the acculturation paradigm which informs the concept of touristic process that was laid out in *chapter 1* to comprehend the entire field of tourism. How does that field look now from the broadest and deepest possible acculturation perspective?

First, consider that tourism involves social contact between people from different cultures or sub-cultures. Like migration, it is an intercultural phenomenon, which should appeal especially to anthropological sensibilities. This contact, which involves leisured travelers and those who serve them, takes place in a context that includes at least two social groups – home and host – and involves various social transactions between the peoples involved. These transactions are primarily of a touristic nature, but may be associated with other transactions (commercial, religious, military, etc.) as well. Thus, in E. M. Forster's *Passage to India*, two of the principal British characters are tourists bent on seeing the "real India", but there also are governors, military men, etc. who are dealing with the Indian hosts. A full anthropological account would take note of these other transactions, as well as the cultures or sub-cultures involved and possible transcultural social structures, in establishing the relevant social context that shapes a particular set of touristic transactions. In *A Passage to India*, that context might be summarily phrased in terms of British imperialism or colonialism.

Second, the social relationship (in this case, touristic) has consequences for the individuals and their cultures or sub-cultures. In *A Passage to India*, both British and Indians have sometimes profound experiences and are changed as a result of the tourists' visit. Looked at from a transactional point of view, the interactions between the parties involved have effects on them, and it is one of the tasks of an anthropology of tourism to assess the personal effects of the touristic transaction, not only on tourists and host peoples, but others such as suppliers of provisions, transport personnel and the like.

Third, the more or less autonomous groups involved in tourism transactions may be seen to be producing (as well as reproducing) cultures in which tourism plays a part. The home culture and associated social structures are generating tourists and tourisms through processes that were discussed in detail in *chapter 4*. The types of tourists they produce ought to be linked, as Butler and Wall (1985:292), remind us, with destination areas and with the character of these areas. On the other side, the host culture produces (or not) those elements needed to host tourism or particular kinds of tourisms. Thus, a particular culture that is as yet uninvolved with tourism may have inns for travelers which can be adapted to serve tourist needs. Or there may be a supply of laborers ready to construct and maintain some kind of tourist infrastructure.

This view that host cultures play an active (productive) role in their own tourism development follows naturally from the basic acculturation paradigm and from work such as that of Young (1977), which was considered in the discussion of the acculturation or development viewpoint in *chapter 2*. Hopefully, it will serve as a useful warning device against easy notions of touristic imperialism among anthropologists and others who may tend to think of hosts as passive objects of outside touristic forces. The truth is that acculturation is always a two-way street even when the power differences between the participants are great. The notion that host cultures are producing (or not) relevant elements for tourism development, which play a smaller or larger role in shaping their destiny, should never be forgotten.

This grand conceptual framework for viewing the entire field of tourism from an anthropological point of view, of course, has to have an historic dimension, which often is referred to by anthropologists in terms of the concept of social reproduction. All of the parties involved in a touristic relationship are in the process of reproducing themselves and their relationships through time. Thus, a particular case of tourism, however it is viewed, always occupies a field that includes a complex that involves a home group, a host group and their relationships, each having its own history. The significant social transactions that combine to produce continuity and change in a specific case of tourism can involve few or many parties and extend near or far. In discussing the related subject of trade, Adams (1974:240) says that those transactions may be "in some respects confined to single communities, in others to multiple groups in time ordered settings, in others to whole regions, in still others to interregional contacts whose historical role is far out of proportion to their limited scale and frequency". It is up to anthropologist observers to situate themselves propitiously for viewing their subject within the changing network of the transactions involved in it.

As an example, consider a plan for tourism development on a small, undeveloped Caribbean island which is a dependency of France. This plan (*Rapport d'Étude*), which has so far not been released to the public nor realized in any actual development, was commissioned by local government from a private research firm in metropolitan France. After exploring the possibilities for tourism development on the island, this firm used its knowledge of the metropolitan tourism market to recommend the development of three different tourism areas. One would concern tourism of sport and rest *en famille*, another, tourism for getting back into shape and another, tourism (more relaxed) for getting away from it all. The plan discusses the various factors that must be considered in developing each area (the lay of the land, the availability of resources, etc.) Insofar as possible, it is suggested that local farming and fishing, already in existence, be adapted to tourist consumption.

If this plan were realized, tourism development on the island would be dictated by an outside research firm which operates on assumptions that are shaped by market demand in the metropole. This firm, as well as the French Government, which ultimately would provide the funds, and most probably, metropolitan construction firms, which would have the responsibility for much of the construction, would then be playing parts in the tourism development on the island.

This proposed plan, if realized, as Lanfant (1980:33) has pointed out for another European-linked scheme in a less developed country, would lead to a form of development that would be heavily determined by outsiders' interests. Research on this tourism development project, therefore, would have to take into account the interests of not only the local government that commissioned the study plan, but also the various outside parties involved (it is unlikely that ordinary citizens of the island would have much of a say). All of these factors (each weighed appropriately) would be included in sketching out the context for this developmental project.

Though the example just given concerns tourism development in a host society, more particularly, a host society in the Third World, it is important to remember that the broad acculturation-oriented scheme being proposed here can be used for analyzing any aspect of tourism wherever it occurs. Questions about any subject the anthropologist chooses to investigate, derived from relevant theory, can be asked and answered from any one of the three anthropological points of view for studying tourism that seems to be appropriate. It is up to the researcher to find the most propitious point of view for asking and answering those questions that guide their research.

Chapter 6

The World of Policy

Tourism Development Among the Mashantucket Pequots

People involved in tourism have their projects, which they formulate and implement, insofar as they can, according to their interests. A look at the development of the fantastically successful casino-resort at Foxwoods in Connecticut by the Mashantucket Pequot Indians, offers an interesting illustration of this process in action. It can also serve as an introduction to the social field in which the applied anthropologist must operate.

Not too long ago, the state reservation on which some Mashantucket Pequot Indians lived in Ledyard, Connecticut wasn't much to look at. A visitor trying to find something of interest saw only a few shacks or trailers and a handful of people who were the resident remainder of a highly acculturated tribe, many of whom had dispersed into the lower reaches of the dominant society beyond. The Pequots at Ledyard earned their living in a variety of jobs, none of them terribly remunerative, and a number of them received welfare payments.

Looking over the general condition of these Indians then, one would have been hard put to envisage that they had any great prospects. Discrimination (doubly difficult for those black Indians among them) had taken its toll; but some of them still were ambitious and astute enough to seize upon opportunities for improving their lives. Following up one of these opportunities, some of these Indians worked out a plan to gain Federal recognition for the tribe and put their reservation and some adjoining land under Federal control. They also sought to obtain restitution from the state for its takeover of their tribal land. These objectives were realized in a negotiated settlement in the 1980s when the Pequots, who had previously filed a law suit against the state, obtained Federal recognition as a tribe and acquired almost a thousand acres of reservation land (under Federal jurisdiction) and the funds to buy more. Though they now had to submit to certain Federal constraints (for example, the necessity of having Federal approval for any land transactions), the tribe was free of state laws, excepting a few such as those pertaining to law enforcement.

Among the state laws not applicable on the reservation were those prohibiting games of chance. This important legal loophole presented the Pequots with an opportunity to get into the gambling business. Their first venture in this direction was a high stakes bingo operation, which was, and continues to be successful. Later, with the passage of the Indian Gaming Regulatory Act (1988) in Washington that formally permitted recognized tribes to engage in gaming, the Pequots moved forward with the construction of a full-scale resort-casino that has become the most profitable in the Western hemisphere. With revenues now approaching one billion dollars, an estimated average of 20,000 visitors each day and close to 10,000 employees, Foxwoods fits easily into the class of Atlantic City and Las Vegas as a gambling resort. A new theater, another hotel, a new building for Bingo, a projected museum, a possible theme park and various transportation projects, are indications that Foxwoods is well on its way into the big time. Another resort is a possibility.

There is another side to development on the reservation which is not noticeable to most visitors. This has to do with the approximately 300 Indians who constitute the Mashantucket Pequot tribe. A community center and sports facilities have been constructed. Housing in the woods for tribal members has been going up. This side of the Foxwoods operation continues to expand, and dispersed members of the tribe have been returning to work or to live out their days in special housing for tribal elders. The shacks and trailers in which the Pequots once lived are difficult to locate these days.

The pace of development at Foxwoods has been so rapid that one might recall the words of an Indonesian Government official to describe it. Referring to development in Indonesia, this official pointed out that his country had been faced with the necessity of having to begin sailing while still building the ship. To assist them in developing Foxwoods, the Indians have resorted to a number of outside specialists (for example, administrators of the casino and of tribal government), outside funding (Chinese Malaysian bankers) and a readily available supply of local labor from an economically depressed part of the state. The key element, however, is the tribal council which is responsive to the 170-odd voting members of the tribe. This seven-member group has been showing some stresses and strains in working out an emerging set of policies for their community and resort.

Some of these policies already are recognizable. There is no question, for example that these Pequots have in mind the development of a big-time gambling resort in the Western mode – a sort of Atlantic City in the woods. They have an even grander idea: the development of a full-scale, family-style resort – a sort of Disneyworld with gambling. This is a fairly typical scenario these days; but on closer inspection, it looks as if the Pequots, as they gain experience in working with each other and the outside world on this project, are putting their own special spin on it. Though it may be too soon to speak easily about Pequot policies for development, one ought to pay attention to indications of these special deviations in trying to figure out where they are headed.

First, the Indians claim that, in contrast to some other gambling resorts, organized crime will be kept out. Second, there seems to be some desire to go easy on the environment, which is, perhaps symbolized by the extraordinary measures being taken to protect the swamp in the center of the reservation; but it is too early to say whether or not these Indians will ultimately line up on the side of sustainable development and stay there. Third, there is an evident effort to recover and preserve an Indian heritage, as, for example in the construction of a museum in the resort complex, the concern for Indian remains on the reservation, the incorporation of Indian-inspired themes in art and architecture, the sponsorship of a native American culture festival and a large contribution to the Smithsonian Institution in Washington for its National Museum of the American Indian. Fourth, the Indians seem to be bent on assuming a share of responsibility in native American affairs that is commensurate with their income. Fifth, with an important lobbying arm in Washington and big contributions to the Democratic party and its candidates, the Pequots have become strong political activists for their own and other Indian causes; and with the emergence of Foxwoods as a major employer, they also are becoming increasingly influential in a region which has been heavily hit by the decline in Federal expenditures for submarine construction and maintenance. Finally, there seems to be a concern with fostering the image of a good neighbor, an image which is being somewhat tarnished by their efforts to buy up surrounding real estate on which they will pay no taxes and their prohibition of unionizing activity (to critics, the Pequots can point out that they pay a portion of their earnings to the state, which redistributes it to local governments, and that their employees are well treated).

The general impression one gets is that these Indians are beginning to feel that they have a significant measure of control over their own destiny, which is, in fact, indicated by their membership in a powerful national lobbying organization, the National Indian Gaming Association, which is working hard and apparently effectively on issues of Indian sovereignty in Washington, the development of a cooperative arrangement with other Connecticut Indian tribes, one of which already is on the way to establishing a casino nearby, and significant contributions to the Democratic Party and its candidates.

All of this suggests that, even though this Indian ship has begun to sail while still being built, policies concerning casino and tribal development have emerged and are being implemented. Unlike what has usually been found by anthropologists in their studies of tourism development among dependent peoples, the Mashantucket Pequots have gained a good deal of control over their own development. It would be a mistake, however, not to see that what has happened on the reservation in Ledyard reflects broader cultural influences and the actions of a number of significant others. First, Foxwoods is obviously a variant of a universal, or near-universal model for casino-resorts in the contemporary world, all of which seem to follow the same general pattern. Second, outside agency has contributed importantly to its development. All of the Federal legislation and court testing that permitted the Indians to prevail in their numerous contests with the state was heavily dependent on outside agency. There also were negotiations with outsiders, as, for example between the tribe, local landowners, the state and (principally) the U.S. Senate, which worked out their transition to a federally recognized tribe living on trust land (the tribe continues to be dependent on Senate actions in this regard). And finally, the hiring of outside professionals necessitated giving up some control over various jobs. There also is routine contact with professionals at other resort-casinos – Indian and non-Indian.

The bottom line, however, continues to be the approval of the tribal council, which gives the impression of operating with increasing assurance in working out the development of the Pequot tribal and resort complexes. There are a number of indications of what amounts to 'Disneyfication' here, among which is the hiring of expert administrators from outside who follow this pattern, but that seems to be the way these Indians want to go.

If policies are taken to refer to the principles on which courses of social action are based, one can increasingly discern a set of Pequot policies for their own development. As the preceding discussion of this particular developmental scenario has made clear, those policies are the outcome of social transactions among interested parties, with the Pequots sometimes (how often?) playing the dominant role. Who are the principal actors in any tourism project and what are their interests? What is their contribution to the setting of policy and to its implementation? For those interested in analyzing and evaluating courses of touristic action, an essentially political analysis of the social processes involved in the formulation and implementation of policy would seem to be essential.

The Social Field of Touristic Action

Looking at any touristic project or event, such as the development of the Pequot Indians and their resort complex, as involving the outcome of give and take between parties bent on advancing their own interests, it is possible to see the social field for such action ranging from simple to complex. In the tradition of those anthropologists or sociologists such as Durkheim (see e.g., Durkheim 1947[1915]), who studied the simplest, most 'primitive' societies in order to more easily understand the universal qualities of this or that social institution, one can explore the social field of touristic action by looking first at tourism in its simplest forms, one of which can be constructed from the reported activities of the aboriginal San of southern Africa.

Speaking of the interband contacts among the G/Wi bushmen (San) not too long ago, Silberbauer (1972:303), points out that most of them took place between allied bands and tended to be kinship structured. The main point of many of these visits was nothing more than pure sociability, the people involved having no utilitarian purpose such as hunting and gathering in mind. The visitors, of course, could have been motivated in a variety of ways. Perhaps they had become temporarily fed up with the social constraints or conflicts in their home band. Perhaps they simply wanted to restore contact with people who were dear to them. If they took off during their sometimes surprisingly abundant leisure time and did not do much in the way of work while away, they (in terms of the argument advanced in this book) would qualify as leisured travelers and hence tourists.

Their tourism, was of the simplest sort, involving only hosts and guests, each having their own interests. Something like this simplest tourism scenario is practiced in the contemporary world by people such as those French urban workers who take their vacations with relatives in the country (see Boyer 1972:50–51). The touristic process may be only a little more complex, among these French, with, perhaps, only a train or bus and their associated personnel intervening.

The agenda of the San guests would have been set by their motivations, which they would be trying to realize as best they could within the constraints provided by their life at home and the contact situation. Among these constraints, surely, must have been the actions of the hosts who, like their guests, may be seen as also bent on maximizing – or optimizing – their own well-being in this particular touristic transaction. They may have wanted to hear the news, chat with someone else, or simply pay back value they, themselves, received on an earlier visit to the band from which their guests originated.

Now let us conduct an hypothetical mental experiment in which we create a situation that may not have occurred in the real world, but could have. Let us suppose that the San hosts in this case had some problems with previous guests. Perhaps too many of them had been showing up. Or they may have overstayed their welcome or behaved badly during their stay. Whatever the nature of the problem, or problems, let us suppose that these particular San hosts had begun thinking about them and talking about them among themselves. In informal give and take, they worked out a collective set of principles or policy regarding guests. Perhaps they arrived at some understanding about when excuses to petitioners for hospitality would begin to flow more freely. Then, they might say that they were too sick or too occupied to entertain guests at that time. Later, perhaps, they might even have worked out specific procedures for carrying out the policy. Who would be the first to implement it, and when? All of this probably would not be as formal as it sounds, but would have been developed in the informal way that was customary among these people.

Suppose we add some outsider who was familiar with San life at that time not too long ago and was on good terms with the people in the hypothetical picture that has been proposed here, as is Lee, one of the anthropologists whose studies of these people have helped to make them famous (see e.g., Lee 1979). He or she could have been called on for help according to the system of generalized reciprocity prevailing among the San and many other hunters and gatherers. This person, not being directly involved in the problem, could help the San hosts sift through the more or less palatable alternatives that were available to them within the context of their culture. Which of these alternatives would they prefer to adopt as a policy for hosting their guests? How would they want to implement it. All of this provides us with a simple form for viewing the world of tourism policy, that is, a world in which people try to figure out a set of principles concerning some touristic project or other, that will enable them to shape it more to their own liking.

Since this is one of those hypothetical experiments of the mind in which we can arrange things fairly freely, why not turn the expert, who was analyzing this particular situation and who was posing various alternatives to the dissatisfied San hosts, into a bona fide applied anthropologist. An outsider, but one who had an understanding of the culture of these particular San, this person would not have been forsaking the quest to understand their ways, but would/could have put that understanding to use in trying to solve a problem for one of the parties in a particular touristic transaction. Having taken sides, as it were, this anthropologist would now be less free as an observer and interpreter and so would have had to monitor additional problems of personal bias in dealing with the issue of unwanted guests posed by the San hosts. This means that the social field framing the investigation would be expanded to include (beside the San parties involved) an anthropologist-actor whose interests extend beyond pure understanding into the world of practical problems.

As pointed out earlier in the book, anthropologists these days are increasingly taking themselves and their procedures into account as factors influencing their studies. With the advent of an applied turn to their investigations, they are faced not only with the necessity of considering their own predilections and those of their subjects, but also those of clients and others in the field in which they are working. As Cleveland (1994:9–10) points out, the values of client and anthropologist do not necessarily coincide. If they do not, there will be additional problems in the transactions between them, all of which must be recognized as factors affecting the picture of the aspect of the touristic process which the anthropologist is constructing. Perhaps, after discussions with the client, the anthropologist will have helped to shape policies; or maybe he or she will decide that the client's aims are intolerable and terminate the relationship. As will be seen in *chapter 8*, there also usually is a middle range of more or less acceptable options in which fruitful applied work can be done.

The mental experiment with parties involved in the simplest form of touristic transaction has served to broach some of the issues that bedevil any anthropologist, especially one who is engaged in applied research. Though it is possible to speak generally about these issues, the specific manner in which they emerge depends on the social field in which the anthropologist is working.

Fields of touristic action in which applied work is carried out are likely to be rather different from the simplest one just discussed. They usually are so much larger and more complex as to be almost of a different order – one that can include a number of peoples and organizations. It is within the largest of such fields, which some have called a world system, that many current questions about tourism have to be considered. Though the relevant actors in different situations where tourism policies are developed and implemented may vary, the ultimate context for comprehending the actions of any party with a touristic interest these days might well be a social field with a global reach. In order not only to understand, but also to act more effectively in carrying out their work, applied anthropologists may find it helpful to think about this big picture and speculate about how it relates to touristic projects with which they are concerned.

The Social Field of Touristic Action in the Contemporary World

In discussing a film by O'Rourke (1987) that depicts the encounter between wealthy Western tourists and Papua New Guineans living along the Sepik River, Errington and Gewertz (1989:275) point out that the film conveys several of anthropology's fundamental messages, one of which is the suggestion "that the conjunction of these (the Papuans') lives and ours – the terms of the interaction between them and us – is strongly affected by the world system". What is this world system about which these anthropologists are speaking so glibly in their comments on one form of international tourism? For some anthropologists and sociologists, it seems to be an established fact to be referred to in the manner of Errington and Gewertz. Other, more cautious scholars are not so sure. They may accept the fact that in the present-day factors beyond the local situation often come into play; but do these factors figure in a social system that extends around the world? One needs to know something about world system theory in order to make up one's mind.

As Shannon (1989:2) has pointed out, world system theory emerged in the 1970s as a Marxist-inspired alternative to modernization theory's interpretation of social change in the contemporary world. According to Wallerstein (1974), who may be the chief spokesman for this point of view (but see also Wolf 1982), this capitalist-based system is supposed to have arisen in the sixteenth century in response to the crisis of feudalism, and though in a period of decline, it continues still. Throughout five hundred years of development, there have been many changes in this system, but its basic structures have remained essentially unchanged.

According to world system theorists, the forces that drive the system are economic of the capitalist sort. Capitalists, who are the central actors, are seen to be constantly trying to accumulate more capital. An important way in which this is accomplished is by squeezing workers who sell their labor to the capitalists who own the means of production. In the contemporary world, the key capitalists are based in core states such as the United States, which now is the most productive of all the competing nation–states in the contemporary world system. Integrated with these core states in an elaborate division of labor are the least productive peripheral states, such as most of Africa, and somewhat more productive semi-peripheral states, such as most of Latin America, which have been primarily agricultural, but may have an increasingly significant industrial component in their economies. It should be remembered that the position of a particular nation–state in the world system is conceived to be in flux. In the early twentieth century, for example, the United States replaced Great Britain as the principal power; and in the 1970s, Japan moved from semi-peripheral status to core.

The world system has been conceived by its proponents as an integrated whole in which essential happenings are economically dictated. What goes on in the nation–states involved is, to a considerable extent, shaped by the nature of their position within the system. On the periphery, development is dependent on relationships with the core, which, because of various inequalities, tend to be exploitive. The rise and fall of core states, on the other hand, is dependent on how 'well' they carry on their relationships with the periphery and semi-periphery. The general view is pessimistic. It is thought that the peripheral states will, because of exploitation, have a hard time being successful in their development efforts; and the whole system is in crisis and doomed to fail. As Wallerstein (1994:10) says, "We are in the midst of a crisis in our historical system, and the violence will much increase before we emerge from it".

Confident assertions like this should not obscure the fact that world system theorists have a long way to go to prove their case. This theoretical approach has excited enthusiasm, but also a good deal of criticism from a number of directions. Among the most telling of these is the theory's failure to include political factors as driving forces, its down-playing of nation–states and what goes on in them, its difficulties with economic cycles, and confusion about the anomalous nature and position of what was once the Soviet Union. As Shannon (1989:179) points out, "Much correction and modification seems to be in order". Still, he argues that world system theory "probably is the most fully elaborated effort in sociology to grapple with the issues of truly large-scale, long-term social change", which are certainly important in the anthropological agenda.

Implications for Policy Research

However one takes world system theory, the value of its overarching conceptions should be apparent to tourism researchers, as well as those who make a living in the so-called tourism industry. That industry is comprised of competing and cooperating business people (the capitalists bent on raising and accumulating capital) and their workers who operate (with more or less intervention from governmental or para-governmental bodies) on local, regional, national and international levels. In some cases (how many?), as Lanfant (1980:23) has pointed out, they may even function in networks that "take no account of national boundaries or territories". How does the tourism industry compare with others?

Douglas Pearce (1989:5) thinks that the parts of the tourism industry tend to have a greater degree of interdependence (destination services are connected with transport, lodging, etc.), that it has a greater number of small, disparate operators (local travel agencies, guest houses, for example), that its market is more differentiated (mass tourism having given way to more specialized tourism) and that, because of the distance between origins and destinations, it covers more space. These qualities, according to Pearce, have certain consequences. For example, overall coordination for political action may be more difficult to accomplish, although one notes the growth of ever-grander, more or less integrated supranational touristic enterprises.

Who are the specific parties involved in this tourism industry? They may be conceived as consisting of governmental and para-governmental bodies such as travel associations, supplier–producers such as transport agencies (airlines, railroads, bus lines, car rentals, etc.), lodging accommodations (hotels, guest houses, campings, etc.), destination services (race tracks, swimming and boating arrangements, bird watching areas, guided tours, etc.), wholesalers (tour operators) and retailers (travel agents), all of whom are organized to promote tourism and to serve tourist needs.

Which aspects of the industry are pertinent for any given tourism operation have to be discovered. For example, the developing Pequot resort-casino mentioned in this chapter is being modeled after other casino-resort operations around the world, but it has its own special character that depends on the Pequot's historical situation, which in this case, includes special relationships with the state and national governments as well as the people living in the surrounding area. The Pequots also are plugged into ethnic networks with other Indians (for political action, cultural restoration, for example) and business associations such as that with other casino-resorts.

Any one of the bodies involved in the tourism industry, or having to do with it, could conduct or commission research to further their own interests. So, a Third World government, such as that of the island of Dominica in the Caribbean, may seek a researcher to assess the carrying capacity of a particular location in the island's interior for a certain kind of ecotourism, or a travel agency may want to find out the social background of its clients and the relationship between that background and their touristic inclinations.

People and organizations outside of the industry, thinking about what might happen in their own backyard, so to speak, also may have a need for research to assess the consequences of different developmental scenarios. For example, one of the functions of the World Tourism Organization is to carry on and make available applied research on tourism to member states. The WTO conducted an impact assessment of tourism in Bhutan (WTO 1986) which suggested that it would be possible to increase the total volume of tourism without altering the carrying capacity of any one location by simply adding other destinations. Anthropologists can bring their own special point of view to bear in all of this research.

Consider how an anthropologist might approach a research question raised by certain clients in the industry, a National Tourism Organization (NTO), Regional Tourism Organization (RTO), or State Tourism Organization (STO). Most countries these days have such organizations. For example, the Republic of Ireland has the Irish National Tourist Board (*Bord Failte*) and a number of regional tourism organizations, which have been studied by Pearce as part of his comparative study of tourist organizations. According to Pearce (1989:9–14), these organizations engage in marketing, visitor servicing, development, planning, research and coordinating and lobbying. It is primarily their research functions that will concern us here, but some attention will be given to coordinating and lobbying.

Pearce (1989:13) points out that (whether they know it or not) "research is vital for these organizations to perform their other functions effectively". It may be conducted either 'in-house' or commissioned from outside. Anthropologists, of course, could bring their expertise to bear in such research, but it is research on the nature of these organizations themselves that looks particularly intriguing to those interested in the social basis of policy decisions. As Pearce (1989:21–22) indicates, an understanding of 'policy' alternatives for an organization can be gained by studying them comparatively, a suggestion that ought to excite anthropologists, for whom cross-cultural comparisons can be a way of life.

Pearce is a geographer, but his approach fits well into the anthropological point of view. An anthropologist would see a tourist organization as having an organizational culture which fits into or adapts to a broader social context (Pearce uses the word "environment"). Looked at comparatively, which helps to establish how a culture actually is constrained by its social setting or context, the policy alternatives for a particular organization are more easily discernable.

Suppose that one of these organizations has a problem and wants to use research to help work out a solution to it. An anthropologist would see that any solution, which, of course, should be effective in dealing with the problem, would have to be compatible with the nature of the organization and its social context or setting. This could be definitively established by comparative study. A number of possible alternatives might be laid out, including extremes which would be difficult for the organization to accept, and further, implement. For example, as Pearce's work makes clear, recommendations involving intervention in local affairs would be less acceptable to the United States Travel and Tourism Agency (USTTA), which is the relevant NTO, than to the New Zealand Department of Tourism because of greater political decentralization in the U.S.A.

The preceding discussion of possible applied anthropological research for governmental tourism organizations has to be taken in an hypothetical vein because no anthropologist has carried out such research. Nor have applied anthropologists done much work for travel agents, tour organizers and others on the demand side of the tourist business. The still modest amount of applied work that anthropologically-oriented researchers have done on the subject of tourism has been almost entirely for clients on the supply side, that is, having to do with destination areas, especially involving the peoples of the less developed world.

Though most Western anthropologists these days still prefer the academic ivory tower, which can provide them a secure base for criticizing the current establishment, applied research has always been a more or less accepted part of the anthropological enterprise. Where tourism is concerned, applied work is not so far along, but the possibilities for such research are enormous. Some of the strictures that ought to guide this work already have been alluded to. Researchers must understand that they are operating in a world in which far-reaching social forces often are at work. Besides comprehending the nature of the problem that is set by the client and the client's organization, the relevant social context has to be identified. In that context are the client and other significant actors, as well as the anthropologist-researcher, each with interests related to some tourism project or event. The tasks facing the applied anthropologist in tourism projects, though they may have a special twist, generally are not unlike those of other applied researchers.

Tasks for Applied Researchers

Applied researchers all consider questions associated with the setting of policy, its implementation and monitoring the realization of a project. Someone (including, possibly, the researchers themselves) wants to know how they and their associates ought to act (the formulation of policy), how to turn thoughts for such action into reality (implementation) and how well such action approaches its intended goals (monitoring). They may be involved in any, or all of these processes. Though scientific considerations are not put aside in such work, some practical problem becomes paramount.

One of the more important problems that has been set for applied researchers on tourism concerns the consequences of tourism for a host society and its environment. The term 'impact assessment' often is used to refer to research of this kind. As pointed out especially in *chapter 2* of this book, this problem has been of much interest to anthropologists and other social scientists who have been studying peoples in peripheral regions of the world.

Though Mathiesen and Wall (1982) refer to a sizable body of impact assessment research, Butler (1993:136) points out that "little of it has been incorporated in what passes for tourism planning". One has to keep in mind, therefore, that the presumed ideal of full cooperation between researcher and client in regard to this problem still is far from being realized. An example is the development plan for the town of Manali in the Kulu Valley in the mountainous Himachal Pradesh area of India, which is referred to by Singh (1989). This plan reflects a policy of preserving the many environmental values of this area, but the formulation of this policy and the plan associated with it seem to have been accomplished without any serious study of the potential impact of tourism development. This is consistent with Inskeep's (1987:120) understated comment that, as far as the environmental implications of tourism are concerned, though considerable knowledge has been gained "it is not utilized as often as it might be".

Ideally, such a plan would offer a number of alternative developmental scenarios, each with its predicted impact, for consideration by policy makers. After studying the situation thoroughly, which involves dealing with sometimes difficult methodological issues associated with assessing the projected consequences of tourism development for host peoples (see *chapter 2*), the anthropologist or other social scientist would lay out various alternatives (possibly ranked according to some desirable criterion) and be available for consultation. For example, numbers and kinds of tourists could be associated with more or less pressure on the environment or existing customs, as in the case of tourist-visitor pressure on Antarctic penguin colonies, which is being monitored by the Argentine Antarctic Institute and the Scott Polar Research Institute (see Acero and Aguirre 1994). The anthropologist might be asked to recommend one of the alternatives, perhaps to install a program for selecting or orienting tourist-visitors for protection of the penguins and their environment. From discussions among policy makers and with anthropological input, a plan for tourism development would evolve. So much for the process of developing policy from an applied anthropologist's point of view.

But as anyone familiar with the field of social action knows, there can be an enormous gap between plans and their realization. Whether or not they have been seriously attended to, plans can gather dust on the shelf or be implemented in such a way as to be unrecognizable. As Pigram (1992:81) says, "Worthwhile policies may be espoused and even formally adopted by management agencies, yet they encounter formidable barriers when attempts are made to translate them into action". The Americans have a way of referring to all this when they say that "talk is cheap". In other cultures, the distinction between thought or talk and its implementation may not be such an issue, but if the people in these cultures want to gain some concrete benefits from tourism, it had better be recognized.

Astute applied anthropologists would include the issue of implementation, as a matter of course, in working out recommendations for policy makers. It would have been taken into account in laying out the relevant social field or context for the tourism project in which they are involved, including the cultural or subcultural background of the significant actors in it. As Smith (1973) points out, conflict and resistance are a normal occurrence in any implementation process, the amount and nature of which is dependent on the nature of the project, how it is perceived by target groups and implementing organizations and the character of the context in which it is supposed to operate. Pigram (1992:82) notes that the successful implementation of a policy depends on whether the people involved are receptive to change, see it as beneficial ("the net outcome will be positive, or at least benign, in the longer term"), and feasible.

Pigram (1992:83) makes a strong case for public participation for both tourism planning and its implementation. He thinks that input by those involved is one of the keys for getting a project implemented successfully, and he goes so far as to say that if residents of a destination area will support a project, tourist-visitors will too. A good deal of applied work in anthropology has been based on the principle that community participation is necessary for the successful accomplishment of any project. For example, the famous Vicos project in Vicos Peru, sponsored by Cornell University and the Peruvian Institute of Indigenous Affairs, sought to identify the sociocultural consequences of introducing a kind of participatory democracy into the developmental process. But the Vicos project ultimately failed in the face of vested interests seeking to maintain a traditional authoritarian way of life, which again raises the issue of conflict between hosts over development projects that was mentioned earlier in the book. An anthropologist does have to wonder about the cross-cultural applicability of any general participatory principle. Nevertheless, the significant actors, their interests and potential power in the implementation process should be routinely considered by applied researchers in any project, including those having to do with tourism.

Sometimes, tourism developers may be aware from the beginning of some attendant difficulties, which are not significant enough to threaten a project; or they may become aware of them only after the project has been implemented. In such cases, an applied researcher could be asked to help mitigate these difficulties after a project was underway or had been completed.

Suppose that in a certain destination area, there is a desire for more tourism and the benefits associated with it, but also a conflicting desire to minimize environmental degradation. In his discussion of the consequences of the advancement of tourism in the Khumbu region of Nepal, which was earlier introduced in the summary case study of Vicanne Adams' work in *chapter 2*, Stevens (1993:410) points to an increased need for fuel by tourist establishments, which has resulted in an alarming destruction of local forests. In order to stem this process, which is considered undesirable by both resident Sherpas and visiting tourists, he recommends the use of available alternative heating and cooking technologies, which Sherpas "have shown much interest in adopting". He also suggests a regional energy arrangement to coordinate the development of such a plan. If realized, this plan would have the potential for mitigating still manageable, but unwanted consequences of development in this destination area. However, though Stevens has been deeply involved in Sherpa life, his recommendations seem not to have been solicited by local powers, which raises serious questions about their possible implementation.

Finally, an applied researcher could be called on to make an assessment of how well some tourism policy was realized. How correct, for example was the prediction of tourism's impact on the host population in the original policy statement? A provision for some such assessment could have been written into the contract of the applied researcher. Such an assessment could provide a check on the prediction and help researchers to revise their procedures. It also might alert those in charge of the development project to problems requiring some kind of mitigation. Butler (1993:149–150) points out that this kind of applied work is practically non-existent where tourism or any other kind of development is concerned.

Summary Case Studies of Applied Tourism Research

Again, in order to bring the preceding discussion in this chapter down to earth, two summary case studies of applied tourism research by anthropologically-oriented scholars will be offered here. One of them, a predictive impact study by Geoffrey Wall (1992), a geographer, deals with the implications of global climatic change for tourism. Though not specifically commissioned by anyone, this cautionary study could be read with profit by anyone involved in the tourist business. Perhaps they will be more seriously interested when, or if the projected 'greenhouse effect' on tourism, considered by Wall, begins to have some palpable influence on their work.

The other case, also an impact study, deals with the consequences of tourism for a specific people, those living on the Pacific islands of Yap. It was carried out by Mark Mansperger (1992) for his Ph.D. dissertation research. During the research, he was invited, though not formally retained by the Yapese Government, to submit his recommendations for tourism development on their islands. He agreed to do this. The summary of his study, as well as that of Wall, is designed to further illuminate the state of applied research on tourism from an anthropological point of view. As usual, it is hoped that these summary case studies will serve as an invitation to look into the original works on which they are based.

Mark Mansperger on Tourism's Impact on Yap

Mansperger, a young American anthropologist with considerable overseas experience, began his college career as an undergraduate in economics. After a brief period as a bank manager in Alaska, he returned to academia for graduate study in anthropology, concentrating first on physical anthropology and later in the sociocultural field. A growing interest in tourism led to a dissertation that dealt with the impact of tourism on small-scale societies, especially on the island of Yap where fieldwork for the dissertation, a revised portion of which is published as *Tourism and Cultural Change in Small-Scale Societies* (1995), took place. Mansperger's study is representative of the bulk of anthropological work on tourism, which has been concerned with the consequences of tourism for host societies, particularly those in the less developed world.

Mansperger's dissertation research took an applied turn with an invitation by the government of Yap to make recommendations for tourism development on their islands. Though still comparatively few in number (about 3,000 mostly young American adventurers in 1990) and easily manageable, tourist arrivals in Yap had tripled from 1985 to 1990. Faced with a severe (negative) imbalance of trade and uncertain about future economic support from the United States (currently obtained through a Compact of Free Association), the Yap Government formulated a policy of developing tourism, as well as fishing and agriculture (note, suggestions about reducing imports were thought to be politically impracticable). Mansperger was asked to assess the current impact of tourism and to recommend a compatible course of tourism development.

The focal points of tourism on Yap have been scuba diving in waters inhabited by huge manta rays, visits to a museum village, and a cultural show or picnic, during which, guests have eaten Yapese food and witnessed traditional dances. All of these activities continue to be Yapese enterprises. Tourists stay at several small hotels with a total of about 75 rooms, all of them Yapese-owned and staffed, or with host families in villages. This means that people of Yap have been the almost exclusive beneficiaries of the small, but increasingly significant income from tourism (Mansperger, however, does not deal with the issue of tourism-related leakage by way of, for example imports necessary to provide amenities for tourists).

In response to questions about tourism by the visiting anthropologist, the people of Yap had some gripes about tourists' insensitivities to their customs, and some thought that tourism was contributing to social conflict and a decline in the custom of sharing, which remains an important component of Yapese culture. But on the whole, Mansperger (1992:12–13) concludes that "the negative influences that tourism is having on village peacefulness, on sharing, and on traditional discipline and tastes are minor". In summary, small-scale tourism development on Yap is seen by the author to have so far brought economic benefits and few social costs to this society.

Is it likely to continue to be that way? Mansperger thinks that the largely benign influence of tourism on Yap has been due mostly to the fact that it has not been a large-scale, foreign-based enterprise. The Yapese have so far resisted the proposals of resort representatives from the United States and Japan to build large-scale resort facilities that include such things as golf courses and beach developments. Were the Yapese to accede to a proposal of this kind, tourism on the island would, according to Mansperger, take on a different quality entirely. Almost certainly, it would become unmanageable in current Yap terms. Accordingly, the anthropologist makes the issue of the possible development of larger, externally-owned facilities central in his recommendations to the Yapese Government. Though he recognizes that financial desperation may drive the Yapese to resort to such development, he hopes that they will be able to avoid it and offers his recommendations with this understanding in mind.

The report is acutely sensitive to the facts of Yapese culture, especially the system of land tenure, which continues today – even after years of colonial interference by Spain, Germany, Japan and lately, the United States – largely in its aboriginal form. This system is composed of a series of hierarchically ordered estates, ownership of which is gained through patrilineal succession. A developer would have to gain control over a number of contiguous estates in a touristically favorable part of Yap (say the island of Map, which has good beaches) in order to develop a large resort complex. What are the possibilities of this occurring?

A possible developmental scenario, according to Mansperger, would go like this: following the passage of a law of eminent domain (pending at the time of the anthropologist's report), outside developers would persuade chiefs and their relatives and loyal neighbors to lease land to them (the sale of land to foreigners being against the law) for (in Yapese terms) large sums of money. Others, not fully aware of the implications of the deal, might be tricked into selling their land to other Yapese who would, in turn, lease it to the foreign developers. Still others might be pressured to go along. This developmental scenario, which would not take place without conflicts over land and other things, would lead to the lease by outsiders of enough land for the construction of one or more large-scale Western-style resorts.

In his recommendations to the government of Yap, Mansperger considers this scenario to be a real possibility. The anthropologist points out that it would result in significant environmental and social costs, including the modification of the system of land tenure, which, as mentioned above, is a key component of Yap culture. The best that the Yapese could hope for if they were to take this developmental road would be to obtain some optimal income and maintain as much control as possible (as, for example, through carefully constructed leasing arrangements) over the development and operation of the resorts. Even so, the income from tourism (and presumably other developmental efforts) might not be great enough to cover the costs of a lifestyle to which the Yapese are becoming accustomed. Then, because of the tourism-induced disintegration of their culture, these people would not have a 'cultural support system' based on the traditional system of land tenure to fall back on.

Better, according to the anthropologist, to gradually develop environmentally and socially friendly tourism on a small scale. This would include expanding diving operations and 'village tourism' in which visitors live in Yapese homes and interact with their hosts in a relaxed and intimate way. Further income from non-diving guests might be generated by tours in glass-bottomed boats and traditional canoes (the museum village and cultural picnic would, presumably, continue in operation, at a somewhat higher level). At the same time, in order to minimize whatever friction exists between the Yapese and tourists, brochures on Yapese etiquette should be made available to visitors. On their side, the Yapese should clean up the trash that has become an eyesore to visitors. Finally, the problem of a growing fresh water shortage should be addressed.

In his recommendations for the development of tourism on Yap, Mansperger is in tune with the spirit of the policy recommendations adopted by the seminar convened in the late 1970s by the World Bank and UNESCO to consider tourism development in developing countries (de Kadt 1979:339–347), which has been mentioned several times in earlier chapters. Included in these recommendations were statements urging cautious and gradual development of small-scale projects, consultations with host populations, an emphasis on unique touristic assets, maximizing host input and the use of host products and personnel, maximizing benefits and minimizing costs on the host side, including the reduction of tourism-induced leakage, striving for an equitable distribution of benefits, cultural awareness programs for both hosts and tourists and rational, integrated planning making use of social scientific expertise.

Though the recruitment of Mansperger by the Yapese Government was ad hoc, this host government was fortunate to have his social scientific input to assist in formulating policies for development. The anthropologist was a sympathetic outsider with an understanding of the culture of Yap, including the state of its tourism enterprise. As a result, the Yap Government received, free of charge, recommendations from a knowledgeable and sympathetic expert, but one wonders what it will do with them.

Mansperger's study of Yap and his recommendations reveal his anthropological background. First, he makes us aware of the international context or setting in which tourism development is taking place. Though he does not use it, the word 'neocolonial' might be appropriate for describing Yapese dependence on the outside, especially the United States. Even though these people are formally independent, they live in the shadow of American influence. We are made aware of how important the Compact of Free Association with the U.S. is for them, as well as of the severe negative balance of trade. One would have wished, however, for an assessment of the role of tourism on this balance sheet. The issue of associated economic leakage to the outside, however induced, looms important in developing countries.

Second, Mansperger attends to the local context in the usual anthropological fashion in his treatment of various aspects of Yap culture, the key element of which (as far as tourism development is concerned) is taken to be the system of land tenure. Small-scale tourism development, according to him, is not likely to undermine this system, while large-scale development certainly will. This anthropologist has come out of the ivory tower far enough to concede that large-scale development may, in effect, be forced on the Yapese people; but he is at pains to point out what he considers to be the pros and cons of this alternative.

Through daily experience in their fieldwork, anthropologists acquire an understanding of the culture of their hosts. Mostly, this understanding is conveyed to the reader in a way that might make many natural scientists wince, which is to say that ethnographic reports are often short on empirical data and weak on method. Mansperger tells us that tourism so far has not had a deleterious effect on this society. He comes to this conclusion on the basis of a survey of Yapese attitudes done before his arrival, his own questioning of the Yapese and his personal observations. Unfortunately, his inquiry appears to be somewhat hit or miss. No clearly stated criteria structure his investigation of this important issue. And finally, there is the bugbear of most anthropological research on tourism development to date: the failure to separate out other kinds of input from tourism in assessing its consequences for some developmental process.

Mansperger does point out that there is a broader Westernizing trend operating, for example through political association and the import of Western goods, but he does not specify in detail how it operates; nor does he lay out the causes and developmental consequences for this host society of different kinds of outside input, only one of which is touristic. This problem is somewhat mitigated by his comparisons with other Micronesian societies in a manner often used in anthropological inquiry and advocated by Pearce, whose work on this subject was cited earlier in this chapter.

Still, one wishes for a more precise ordering of accounts in this particular developmental agenda. One is inclined to believe that Mansperger is in touch with the norms of Yapese culture and to accept his statements about them. However, where sociocultural changes and their causes are concerned, his explanations are somewhat vague. It would be nice to have an empirically sounder re-study from this anthropologist that treats the change process in more detail.

Though the hard-headed scientist may have reservations about Mansperger's analytical work, one can be much more sanguine about his recommendations for tourism development, which certainly are in line with those of experts for societies of this kind. As those experts convened by the World Bank and UNESCO pointed out (de Kadt 1979:339), tourism, if carefully planned, "can make a substantial contribution to the economic and social development of many countries". Mansperger's recommendations show how this might be accomplished on Yap. If that does not happen and the Yapese are forced down the alternate road to large-scale tourism, it will not be because they have not been warned by their applied anthropological friend who is keenly aware of their ways and the possibilities of developing them beneficially through tourism.

How much influence have Mansperger's recommendations had on Yapese policies regarding tourism and other development? And what of the implementation of these policies? In the best of all possible worlds, from an applied point of view, the anthropologist would assist not only in the formulation of policies, but also in implementing them and making assessments afterwards. Nothing like this has happened with Mansperger's carefully-thought-through recommendations. Since the time he sent them to Yap authorities, he has heard absolutely nothing from them. Letters of inquiry have gone unanswered. He says (in a personal communication), "My feeling is that they did read the report, but I have no idea of its impact".

Geoffrey Wall on the 'Greenhouse Effect' and Tourism

Geoffrey Wall is a Canadian geographer with broad and varied interests. His special fields include tourism, leisure activities, and resource management. Currently, his international experience is being extended through work in Indonesia, particularly in the Bali Sustainable Development Project, which will be referred to in detail in the next chapter. Wall's extensive applied work has been mostly in the service of governmental and para-governmental clients ranging from local to international. Working out of his base at the Canadian University of Waterloo, he has gained an international reputation, which includes charter membership in the International Academy for the Study of Tourism.

The study to be summarized here, *Tourism Alternatives in an Era of Global Climatic Change* (1992), is one instance of Wall's applied work on the relationship between climate and tourism in a variety of destination areas. This impact study, which is derived from earlier applied work, pays attention to several destination areas in Canada, but only after considering the effect of a possible climate change on tourism around the world. Here is one of those studies in which scholars attempt to fathom the future. That future, according to Wall, will be subject to a significant climatic event which will affect many human activities, including tourism.

If the best scientific projections are to be believed, humankind will soon be forced to adapt to the most significant climatic change since the glaciers last ceased their ebb and flow thousands of years ago. Apparently already under way, a human-generated 'greenhouse effect', which involves the warming of the earth's atmosphere through the accumulation of carbon dioxide, is examined for its effect on tourism worldwide, as well as in specific locations.

In his discussion of the "climate–tourism interface", Geoffrey Wall points out that though climate imposes constraints on tourism and other human activities, the nature of these constraints often is, with the exception of those on life, itself, activity specific. So, a different type of climate is necessary for different outdoor activities such as skiing, swimming, viewing autumn foliage, and the like. If the 'greenhouse effect' is fully realized, as projected by experts such as those convened at the 1985 Villach meeting sponsored by the International Council of Scientific Unions and the United Nations, certain consequences for tourism should follow. It is expected that warming will be significant, but variable around the globe, with the warming being most significant in the highest latitudes (projections about precipitation are much more questionable). One projected environmental consequence of this would be an increase in sea level around the world of between an estimated 20–140 cm. Another would be the decline in snow cover in the higher latitudes.

Such developments would mean that those tourist destinations that depend on certain kinds of natural resources will be at greater risk. Shoreline areas such as at Venice, Italy and the beaches of the Mediterranean, could be substantially altered or even obliterated. Destination areas, such as in the Caribbean, which have come to depend heavily on beach tourism, could be faced with the elimination of a major source of their income. Wall points out that the temperature increase also could cause many tourists to change their destinations to cooler climes in higher latitudes or altitudes.

After discussing the rather disquieting implications of global warming for tourist destination areas generally, the author turns his attention to particular locations in Canada, with which he has had considerable experience. Some of these locations are oriented towards eco- or nature-tourism. Others feature skiing. Still others are mostly camping sites. Wall's analysis of two skiing areas in Ontario is instructive. The more northern of these areas in the Lakehead region has a longer skiing season than does the more southerly area near Georgian Bay. 'Better' and 'worse' case warming scenarios are developed for each. In the 'worse' case, skiing would be virtually wiped out in the South Georgian Bay area and reduced at Lakehead, but because the key vacation periods over Christmas and during the mid-February university/college break would not have significantly less snow cover, a viable downhill skiing industry could be maintained there (Lakehead). In the 'better' case, the ski season would be reduced in both areas, but the prospects for the Georgian Bay area would be so unreliable as to seriously threaten the skiing industry there. Wall concludes that, in this case, Georgian Bay and more southerly regions such as the Laurentians probably would be eliminated as skiing destinations, but the more northerly areas could continue to operate.

One hopes that those scientists, whose projections of global warming Wall uses, are wrong. Some people may take comfort in the fact that not all experts are in agreement on the subject. Others, believing the worst, may be doing what they can to retard or stop global warming or simply accepting it as a fact of life. Wall does recognize that the projections of his experts may not be realized. He also recognizes that other changes, not controlled in his scenarios, could have significant effects on tourism worldwide and at specific locations. Still, one gets the impression from Wall's work that the tourism industry had better take the possibility of global warming seriously and consider alternatives to current practices.

The article by Wall on the possible consequences for tourism of global warming is written with the authority that comes from extensive experience doing impact studies for a variety of clients. Mansperger's recommendations, on the other hand, do not convey the kind of assurance that comes from being a practiced participant in the field of tourism policy development. He was, after all, only on the verge of receiving his Ph.D., had no history of applied work, and was not formally retained by the Yap Government to do the study. In many ways, his case is representative of applied anthropological work on tourism. That work still is in its formative stage; and though, as should be clear from the earlier part of this chapter, anthropologists studying tourism have much to offer policy makers, their contributions have been comparatively insignificant. But Wall's work and that of the geographer, Pearce, demonstrate that other social scientists with an anthropological orientation are already seriously involved in shaping tourism policies.

Does it matter who does this work? Certainly, there is a good deal of overlap in the approaches of geography, for example, and anthropology, but anthropology does have a distinctive vision that can be of use to policy makers. That vision, besides paying attention to the sometimes very broad context that shapes touristic events, also offers something of an insider's view of the human beings involved in them – something that is missing from Wall's discussion. In their consideration of the long sweep of human history and pre-history, some anthropologists also have dealt with environmental happenings no less significant than the projected global warming considered by Wall and the adaptations of humans to them. Others, in their investigations of the sometimes wrenching changes involved in development have considered the meaning to the humans involved of these changes in their lives. Accordingly, anthropologists are able to draw on a background of relevant experience in addressing human problems of whatever nature and scope. But as far as tourism is concerned, anthropologists have done comparatively little to solve the human problems it poses. Why is this so?

Some might say that it is simply a matter of anthropology discovering tourism comparatively late in the game, and it is only a matter of time before anthropologists find their way into the field of tourism policies, their implementation, mitigation and assessment. Geographers, they might argue, have a longer history of interest in tourism and are more familiar with the touristic scene. Those, like Wall, use that greater familiarity to move easily around in the world of tourism policy. Others, not so sanguine about the prospect of anthropologists coming to be at home in this world, may point to a certain anthropological orientation that could prevent it.

That orientation, which has come naturally to many sociocultural anthropologists, especially since the time of World War II, has had a substantial anti-establishment component, which is in favor of the have-nots of the world, who often are conceived of as being exploited by more developed countries. Accordingly, any project designed to make any exploitive system work better would be viewed by these anthropologists with reservations. They would not be so interested in trafficking with anyone in an allegedly exploitive industry who might seek out their services.

Possibly, as in the case of tourism development in the less developed world, there are acceptable tourism projects designed to help the deprived make out better under the system. But going to work for one of the international tourism conglomerates such as an airline, cruise line, or resort chain would almost certainly be regarded by many anthropologists as a kind of selling out. Even work for a Third World government such as that of Yap, which is deeply dependent on the United States, might be construed by some anthropologists in this way. They would prefer to keep their distance in some university ivory tower from where it is easier to mount their critical analyses of the world as it is.

One senses in the work of Wall and a good many other social scientists less severe compunctions against helping the members of some establishment with their problems. Certainly, tourism researchers working for governments with command economies such as China (see RDCNTA 1989) do not reveal such a super-critical attitude. And there may be more Western anthropologists than is generally realized who will follow the policy trail wherever it leads. Indeed, the author recalls a crowded conference at a meeting of the American Anthropological Association in the late 1970s in which a representative of the World Bank discussed its plans for aiding tourism development. It was evident that many of those attending had been attracted by the possibility of World Bank support for applied research on the subject. Applied anthropology continues to make contributions not only to the solution of human problems, but also to the discipline of anthropology itself. There would seem to be no reason why this also should not be so where tourism research is concerned.

Chapter 7

Towards More Sustainable Tourism Development?

The title of the book by Turner and Ash conjures up vivid images. *The Golden Hordes* may bring to mind a shining swarm of invaders bent on massacre and laying waste to the countryside. The authors, themselves (1975:11), are thinking of masses of barbarians who, like those of the Tatar Empire, threatened the urban civilizations of Europe. But the 'barbarians' they are referring to, while also destructive in their own way, bring with them something of value. Indeed, this is the way that Turner and Ash want to portray the advance of mass tourism into destination areas on the periphery of the contemporary Western World. On the one hand, these tourists are a source of precious foreign exchange that can contribute to the local income, but on the other, according to Turner and Ash (1975:15), their arrival results in "the systematic destruction of everything that is beautiful in the world".

By this time, readers should be familiar with this point of view, which, like much early anthropological work on tourism, had a cautionary or debunking quality. It warned people in destination areas about being too enthusiastic about the prospects of tourism development. Anthropologists, though entitled to some caution about establishment-dictated events, ought to have acquired a broader view by now. Certainly, tourism can have negative – even catastrophic – consequences for a host people, but there can be a positive side too. It depends on the kind of tourism and the nature of the acculturation situation in which it operates.

As *chapter 6* made clear, applied research on the subject by anthropologically-oriented researchers has mostly taken the form of impact analysis which contributes to an awareness of the consequences of whatever form of tourism is being developed. In such research, as in other analyses of development, alternatives to mainstream tourism, which are ecologically sound and respectful of the needs of all involved, are being explored. In this regard, the concept of sustainability has come into vogue. Sharachchandra Lélé (1991) says that "SD (Sustainable Development) has become the watchword for international aid agencies, the jargon of development planners, the theme of conferences and learned papers, and the slogan of developmental and environmental activists". It was given a special cachet in the famous 'Brundtland' report produced for the United Nations by the World Commission on Environment and Development (WCED 1987). And de Kadt (1992:56) thinks that making it the focus of alternative tourism development may be the most productive way to move forward in the policy area.

There could be a danger, however, in such widespread and easy acceptance. The concept may have acquired such a taken-for-granted quality that people are no longer concerned with what it means. As a consequence, its function as a tool for scientific analysis may not be critically considered. It turns out that the term sustainable development has a number of meanings, each of which grew out of certain developmental concerns. Some reference to these concerns ought to help develop a critical understanding of the term and to use it more wisely. This chapter aims at providing this critical understanding and shows how the concept, if properly employed, can help in projects designed to further an alter native to the kind of tourism criticized by Turner and Ash.

A Little History

The concept of sustainable development can be related to the conscious concern with social development that emerged a half century or more ago, particularly in the less developed or non-Western world. In the countries or places that make up this part of the globe, people tend to have a lower standard of material well-being than exists in the West. In fact, as indicated by sometimes calamitous death rates and excruciating poverty, a good many of them actually may have trouble surviving. By the middle of the twentieth century, most of these countries had become formally independent of Western colonial rule, but they still were likely to be in a state of economic or other dependence and to show the marks of years of colonial domination.

Now more than ever aware of the deficiencies in their lives, people in this less developed world began to think about, and plan changes that would eliminate these deficiencies, that is, they began to think about what has come to be called development. Though development ultimately may be thought to involve change in any socially desirable direction, more often than not, it has been conceived in economic terms and to involve increasing productivity – often with assistance from more industrialized nations – as demonstrated, for example, by an increase in the gross national or domestic product. Though there are a number of bottlenecks to economic development, among which are inadequate natural resources, low technological skills, deficient institutions, low demand for products, a lack of capital, and continued foreign exploitation, most of those countries in the less developed part of the world have embarked, for better or worse, on some kind of economic development.

In some cases, this economic development has been pursued at all costs, with harmful consequences such as the kind of degradation of the environment that earlier appeared during industrialization in the West and now is all too evident in Eastern Europe and certain parts of Asia. There also is a general tendency for benefits to flow to the rich and powerful in and outside of a country and a drift towards dependency on outside powers. In other cases, though, conscious attempts have been made to steer development in more desirable directions such as safe-guarding the environment.

For a less developed country to embark on a course of development these days, there usually is some kind of Western input, which can include tourism. Whatever the scheme it employs – whether capitalist, socialist, Islamic, or something else – it is more likely to address, first, its economic condition rather than its social or environmental conditions. An example of such development, which is noteworthy for its attention to tourism-induced productivity, but also to social issues, is reported by Boissevain (1978) from Malta where early returns on the effort to develop tourism in a socially responsible way showed that it, indeed, had contributed to social equality, heightened Maltese cohesiveness, helped to maintain traditions, and decreased dependence on the outside (later returns, however, indi-cate a dissatisfaction with economic benefits and a growing dependency on the United Kingdom, which is the source of most of its tourists. See Oglethorpe 1984).

Though there are fervent believers in some developmental scheme or other in both developing and developed countries, it is not so easy to find instances of success. One can point to countries in Asia such as Taiwan and South Korea, where economic success, at least, is quite evident; but there are enough 'bad' examples not only on the economic side, but also on the social and environmental side to suggest caution regarding any developmental scheme. Moreover, we still have a way to go in under-standing the developmental process. As de Kadt (1992:48) points out, no existing economic theory "has given a satisfactory overall explanation of the process of development, or of its 'absence'". The watchword for the moment would appear to be to think small and apply any general theory of development with caution. And, of course, it might be wise to keep in mind the old American aphorism, presumably universal in its application, that there is no free lunch.

Alternative development, that is, development, which tends to put a little less emphasis on economic productivity as compared with social and environmental goals, has been thought of in various ways, and labels such as 'soft', 'green', 'participatory', 'humane', 'non-exploitive', 'small-scale' and 'responsible' have been used to refer to it. Increasingly, one hears the word 'sustainability' used as a kind of shorthand way of referring to such alternatives.

As Sharachchandra Lélé (1991:610) points out, the term sustainable development began to be used in the 1980s as a new way of expressing a conscious concern for the conservation of nature, a concern which, in the West, goes back a century or two, but has become particularly salient in the latter part of twentieth century. The most radical of the environmentalists (sometimes referred to as deep ecologists) have argued for nothing less than a complete reorganization of society and its relationship to the environment, with the possibility of 'no' economic growth as a real option. A recent influential book on this subject by Roderick Nash (1989) is entitled *The Rights of Nature*, which are taken to be at least as important as human rights. Such a position, of course, contradicts that of mainstream developmentalists who think of society and the environment more as resources for development, not as things that need to be protected.

Most environmentalists, however, have made some sort of more or less conflictful accommodation with developmentalists, and vice versa. It is with the nature of this conflict-laden accommodation, especially regarding tourism development, that this chapter is concerned. The question will be asked how anthropologists can use the concept of sustainable development to further their work. To begin, consider one aspect of this, which involves protection of the environment.

Environmentalism in the United States

A brief history of environmental consciousness and associated actions in the United States can serve as an introduction to the issues associated with the concept of sustainable development. As was pointed out in *chapter 4*, Americans have been thinking about nature for a long time. Sale (1993:5) points out that in the first half of the nineteenth century artists and writers of the Romantic and Transcendental movements had already laid an environmental foundation to build upon. If Roderick Nash is to be believed, this concern spread to some extent in opposition to the advance of industrial-commercial enterprise. Dunlap and Mertig (1992:1–2) think that contemporary environmentalism in America is historically related to the conservation movement "that emerged in the late nineteenth century in reaction to reckless exploitation of our natural resources". Then, people like Gifford Pinchot and John Muir began to argue for the preservation of nature for its own sake, the first national parks were created, government agencies concerned with nature, such as the U.S. Forest Service, were established, and the first conservation organizations, such as the Sierra Club and the National Audubon Society, came into being.

The American environmental 'movement' was given a great boost in 1962 by Rachel Carson's (1962) book, *The Silent Spring*, which treated the wide-ranging effects of pesticides on the natural environment and on human beings, and by the first Earth Day in 1970 in which some 20 million Americans are said to have participated. During this period, additional impetus was given to environmental concerns by protesters against the Vietnam War whose analyses of the failings of American society included its treatment of the environment, and the emergence of a larger, college-trained, white collar, middle class interested in quality of life issues.

According to Dunlap's (1992) study of American public opinion, this initial burst of enthusiasm was followed by a modest decline in commitment to environmental issues through the 1970s, but another significant increase (perhaps in opposition to the Reagan administration policies) occurred in the 1980s. Dunlap (1992:110–112) says that "by 1989 majorities of the public (often around two-thirds) were rating most of the (environmental) problems as large threats to the environment as well as to themselves". He concludes that the American public now is aware of environmental degradation and wants to do something about it (they also, of course, are aware of the facts of economic recession, which may cause some diminution of environmental concern, as may be indicated by a recent decline in commitment to core environmental organizations).

Though the American Government and big business have become more responsive to environmental issues, it seems that the interests of the developmentalists often have prevailed. True, developmentalists now must contend with increasingly restrictive environmental legislation, as well as a concerned public opinion, and make more or less significant compromises in their undertakings, but they seem to be better informed politically, have greater financial backing than their environmental opponents, and are more efficient in their operations; and by playing their 'more jobs' card when needed, developers often seem to gain the upper hand. Environmentalists have made significant gains, it is true, but even so, most Americans now think that the quality of their environment is deteriorating (Dunlap 1992:112). McClosky (1992:86–87) believes that, as far as American environmentalists are concerned, the issue no longer is how to gain public support, but how to produce results.

Not too long ago, most of the people in the American environmental 'movement' had put their faith in a policy of developing a responsive government. The concentration was on political-legal action to bring about government regulatory programs, say, concerning toxic wastes, water quality, soil erosion, or air quality, that was sympathetic to their views. Thus, a mainstream environmental organization such as the Sierra

Club would work to get its candidates elected and lobby those candidates and others for appropriate legislation. But even when such legislation was in place and appropriate policies formulated, there has been an annoying lack of follow-through or implementation, which can be traced to bureaucratic intractability, countervailing political pressures (especially in the case of the Reagan administration), hostile administrators, insufficient personnel, or simply inertia. Moreover, even putting the Reagan administration aside, no American presidency seems to have been willing to draw a line in the sand, for whatever reason, against the forces arrayed in favor of economic development, which are currently gaining new influence with the rise of Republican representation in the U. S. Congress. This, according to McClosky (1992:86), has led to grassroots dissatisfaction and a search for alternative strategies.

In order to keep the record straight, one ought to remember that American developmentalists are not now as they used to be in the old days when industrialists such as Andrew Carnegie could do pretty much whatever they wanted. As a matter of course, they must now take into account increasingly restrictive environmental legislation which the new Republican majority in the U.S. Congress is currently seeking to dismantle. Environmental impact statements have to be filed, environmental regulations attended to, and environmental glossings given for public consumption; but, as de Kadt (1992:57) points out, it still is rare for the costs that environmentalists worry about to be internalized by a developmentalist. As he says, "often these costs will not be 'calculated' by the users unless external authority forces them to do so".

American environmentalism is, of course, only one of many environmental movements operating in the world these days (a particularly important point of environmental activism centers on the Green Party in Germany). The American case has been singled out for consideration here because the main thrust of environmentalism has come from the industrialized world and because America is the leading industrial nation. One should keep in mind, though, that American environmentalism, as well as other national environmentalisms, are becoming increasingly internationalized. As Caldwell (1992:73) points out, "the globalizing of U.S. environmentalism is merely a particular case of the globalizing of environmentalism everywhere". Even as the forces arrayed in favor of economic development have become increasingly internationalized, so have those seeking to protect the environment.

Among the issues being addressed by governmental and non-governmental environmental organizations on an international level are nuclear proliferation, international transport of hazardous materials, changes in the ozone layer, global climate change, threats to wildlife, and destruction of tropical rain forests. These issues have been given a certain legitimacy by conferences between heads of state and within frameworks provided by the United Nations. The Brundtland report, prepared by the UN-sponsored World Commission on Environment and Development (WCED 1987), is the result of one of those conferences.

From all of the foregoing, it should be clear that the question of development has increasingly come to the fore in today's world. It is hard to find a country that is not officially in favor of some version of it. But economic development, which seems to be universally desired and which is necessary to raise a standard of living, does not come without costs. The various environmental 'movements' which have been concerned not only with the natural, but also the social environments of human societies, have had a lot to do with making us consider these costs.

We have become more aware of the sometimes catastrophic effects of mainstream economic development, and many of us are willing to entertain – more or less seriously – alternatives to such development. That, when push comes to shove, we may still tend to favor a measure favoring greater productivity (and jobs) over one that offers an alternative with greater environmental and social protection should not detract from the growing cultural significance of the latter, which, increasingly, is being viewed under the rubric of sustainable development.

The Meanings of Sustainable Development

According to the Brundtland report (WCED 1987:43), "Sustainable development is development that meets the needs of the present without compromising the ability of future generations to meet their own needs". In the simplest interpretation of this phrasing, a people cannot project a course of development that goes beyond their own needs *and* those of future generations. This cautionary statement, which was elaborated and acted upon at the United Nations Conference on Environment and Development at Rio de Janeiro in 1992, raises a host of questions about human needs and the resources to satisfy them. It also suggests that any developmental program must take into consideration a broad range of human and environmental factors now and in the future. One might interpret this to mean that human needs must be satisfied over a long course, but so, also must the 'needs' of the environment with which humans are intimately connected.

Lélé (1991:609) points out the close connection between the concept of sustainability and environmental concerns. He believes that for most people the term refers primarily to maintaining an environment that will support human life "at some specified level of well-being" now and in the future, which is one way in which the Brundtland statement can be taken. Interpreted in this manner, sustainability comes very close to the popular idea of carrying capacity, which, according to Dewar (1984:601), "reflects the commonsense notion that a limit on resources implies a limit on the number of consumers". As an anthropologist is likely to point out however, what is a limit for one people, is not necessarily a limit for another. Certainly, there are basic needs to be satisfied, but what about culturally derived wants? They vary greatly from one society to another, as does the effectiveness of utilizing the environment. This suggests that sociocultural factors needs to be considered in any discussion of carrying capacity or sustainability; and in the Brundtland report, which Lélé (1991:611) believes to be representative of the thinking of most advocates of sustainable development these days, such a view is clearly evident.

In order to "meet the needs of present and future generations", the report argues for the kind of economic growth that will be sustainable. Some kind of economic growth is necessary because millions of humans are in want, and, indeed, an increase in productivity may be necessary to prevent the poverty-stricken from pillaging their environment; but such growth cannot exceed the tolerance of factors on which human life depends. The environment must be used more intelligently by doing such things as developing more appropriate technology. Human needs also must be limited by various strategies including population control. And the fruits of economic growth must be spread around more evenly by the reorganization of economic systems and by having the currently deprived take a more meaningful part in the developmental process.

It is clear from all this that sustainable development involves a number of interlinked goals, which its advocates, while perhaps accepting the fact that significant trade-offs will be necessary, believe can be made consonant with one another. Lélé (1991:616–618) points out a number of problems with this facile view. For example, he raises questions about whether social equity necessarily leads to environmental sustainability, or vice versa, whether economic growth always can be made consistent with both sustainability and the removal of poverty, etc. In regard to such difficult issues, Lélé (1991:615) argues that the proponents of sustainable development have adopted a "narrow-minded, quick-fix, deceptive approach" which may involve an assumption of being able to have your cake and eat it too.

All of this suggests that the concept of sustainable development provides a kind of catch-all or umbrella term for many of the inherited concerns that have stirred alternative developmentalists, including environmentalists, in their opposition to mainstream development. These alternatives, according to Dorcey (1991:5), include, at a minimum, maintaining ecological integrity and diversity, meeting basic human needs, keeping options open for future generations, reducing injustice, and increasing self-determination. However, not only are the various objectives often ambiguous, but there is a real question whether they do, in fact, constitute a single, coherent and realizable package.

To begin to sort things out, it might be wise to begin with the issue of ecological soundness and ask what is to be sustained in what regard. Is it the environment, people, or sociocultural arrangements? Then, one could go on to ask other crucial questions about the potential beneficiaries of such environmentally sustainable development. Is it a few conservationists, everyone in the host society (the issue of distributive justice), everyone in subsequent generations (intergenerational justice), etc.? Such a course of hard questioning and subsequent analysis would have the salutary effect of demystifying what may be turning out to be a sacralized ideology.

The Tourism Connection

In the early 1960s tourism already was being recognized as a possible contributory factor in the economic development of less developed countries (see e.g., United Nations 1963). This recognition can be associated with the study of tourism development in these countries, much of it funded by outside agencies such as the World Bank. Associated with such development was an increase in tourist arrivals who were supposed to bring economic benefits with them. According to Goldfarb (1989:6), "Developing countries' share (of international tourist arrivals) was 6.5% in 1960, increasing to 11.1% in 1970 and 16.6% in 1980". There was then (and continues to be) a good deal of enthusiasm for the use of tourism as a developmental tool, a position which corresponds to Jafari's 'advocacy platform'. As pointed out for Gambian tourism development in *chapter 2,* the Gambian Government and outside experts from the U.N. and the World Bank appear to have shared in this enthusiasm.

But already in the 1960s, cautionary notes concerning tourism's developmental potential were beginning to be heard. One theme that has run through this literature, a good deal of which has been Marxist inspired, is that input from the more developed world, instead of helping a developing country, actually may, on balance, harm it. So, the construction of a big oil refinery with foreign aid can result in damage to the environment and the fabric of a society, not to mention the creation of an exploitive dependence on outside parties. Though tourism development may not initially seem to have such a destructive potential, it can be equally damaging in its own way.

An early indication of an awareness of this is to be found in the vituperative writing of the Martiniquen-Algerian psychiatrist and developmental theorist, Franz Fanon (1967:123–124), who saw tourist resorts in the Third World such as Havana and Acapulco as entry points for an imperialism that would eventually turn their countries into "brothels of Europe". An echo of this view is to be found in the socialist journal *Dollars and Sense* (1978:15), which argues that "luxury tourist accommodations run by TWA don't promise any more in the way of adequate economic development than do copper mines run by Kennecott or Mustang assembly plants run by Ford".

Fanon's and other radicals' solution for this problem, which is intimately associated with what has come to be known as Dependency Theory, is a revolution that would completely cut the ties of dependency of Third World peoples on the Developed World. Others, while perhaps posing the problem in equally stark terms, propose less radical alternatives to mainstream tourism development. For example, Srisang (1989:119–121), writing for *The Ecumenical Coalition on Third World Tourism*, which was mentioned earlier as one of a number of international organizations bent on promoting more responsible tourism in the Third World, argues that in the current world crisis peoples in the Third World are suffering the most; and contrary to the claims of its promoters, tourism not only is not helping to resolve this crisis, but actually is worsening it for these peoples. He says that the coalition, in aiming to tell the truth about the negative impacts of current tourism and to help to transform it, seeks "to participate in the search for a just, participatory, and sustainable society". This view approximates that of Catholic liberation theologists working through Base Christian Communities in Latin America. That such views can be more than empty talk from the discontented is indicated by the effective political-legal action against mainstream tourism development in Goa by a militant local group calling itself the Jagrut Goencaranchi Fauz (see Lea 1993:708–710), which succeeded in halting hotel construction in a desirable tourist area and even causing the demolition of a number of hotels that were being constructed illegally.

An even more impressive indication of this countertrend is to be found in the operations of the World Tourism Organization, where sustainable tourism development has become an integral part of its policy. This organization, initially a flag-waving promoter of tourism, generally, has become "greened" to the extent that conferences on sustainable tourism are routine and advice to clients on tourism development normally includes guidelines for sustainability. The ideology of the organization in this regard is particularly apparent in its publication *Sustainable Tourism Development: Guide for Local Planners* (WTO 1993).

The *Guide*, after recognizing its indebtedness to the Brundtland report and to the Rio conference of the United Nations, proceeds to look at tourism as a global phenomenon. It begins with a dose of reality. Tourism development is not for everyone. Not all societies have appropriate attractions and activities for tourists nor the resources to handle them (WTO 1993:5–7). But some societies can benefit economically and socially from sustainable tourism development if three principles of sustainable development are followed. These principles (WTO 1993:10) concern ecological sustainability, social and cultural sustainability, and economic sustainability for present and future generations.

The *Guide* proposes that the various groups with an interest in tourism development – the tourism industry, environmentalists, and local communities – work together as partners for sustainable tourism development. In the discussion of this subject (WTO 1993:16–38), the *Guide* makes liberal use of the concept of carrying capacity, specifies the interests of the parties involved, and emphasizes that these interests must be reconciled according to the framework provided by the principles of sustainability in order that "an improved quality of life can be achieved".

Subsequent chapters in the *Guide* concern the preparation of plans for developments, which are viewed as integrated wholes. There follow guidelines for the implementation of these plans (including marketing), in which the activities of public and private agencies and their coordination are laid out, the management of environmental and socioeconomic impacts, so as to increase benefits and minimize costs, and finally, the management of the tourism sector, which stress the necessity of continuous monitoring and mitigation.

From this brief review, it is clear that the authors of the *Guide* believe that sustainable tourism development is not only feasible, but attainable in the contemporary world. They believe that in certain situations (how many?) all involved can, by working together, benefit. The fruits of such development will not come automatically, but through hard-nosed, intelligent planning and resolute implementation, monitoring, and mitigation. Of course, the values of mainstream development as it has been practiced in the past, will have to be transformed in line with the principles of sustainability, but the *Guide* is basically optimistic in this regard. Perhaps the authors feel that by looking over the havoc that often has been created by mainstream tourism development, developmentalists will come to feel that the present way of doing things has to be changed for some alternative. They seem to be thinking that it is worth giving this one a try, especially if funds are available for it, or tourists and local communities are beginning to insist on it.

A recent presentation by the President of the International Federation of Tour Operators, Martin Brackenbury (1993:17), gives some cause for optimism in this regard. He says, "But grow as it may, there is no excuse for any style of tourism to inflict damage to the place in which it operates. Enough is now known to require any tourist development, new or existing, to abide by the principles of sustainability".

What of the developers and managers of the tourism industry who are most likely to hold the key to tourism development? After all, if they can't be brought round to acting according to the principles of sustainability, rather than those associated with short-sighted material self-interest, the tourism they create and maintain won't become sustainable.

Here, de Kadt's discussion (1992:61–75) of the forces arrayed against sustainable tourism development is on the mark. First, take the disparate members of the industry – the entrepreneurs, bureaucrats, bankers, managers, and the like – most of whom so far have not internalized principles of sustainability in their calculations. The environment and society do have their values that can be factored into any tourism development project, but members of the tourism industry often to not appreciate their full value. Speaking only about marina development in some British estuaries, Stabler (1992:369) says that "if total environmental value is substantial, many marina developments, hitherto considered commercially viable, would no longer be so".

What would make tourism developers put a greater value on the environment? First of all, there is public opinion, which – especially in the more developed countries – is increasingly in favor of sustainable values and might turn sour on them if these values were openly violated. Second, there is the increasingly constraining network of laws that favor the sustainable side. Industry members, with their money and political know-how may still be able to work around these laws, but they no longer can be disregarded. Finally, there are economic considerations, which, sooner or later, must be attended to. It has been demonstrated that in some cases (but not all), the environmentally and socially friendly alternative actually turns out to be 'economically' friendly. So, Brackenbury (1993:17), in another part of his remarkable statement, says, "Our clients, the world's travelers, will simply no longer go to places that are polluted, where landscape has been lost. The end of environment is the end of tourism. Full stop".

Echoes of this sentiment may be found in a resort like Zermatt in Switzerland, where the construction of certain kinds of hotels in certain places is not permitted because they will interfere with views of the Matterhorn from existing hotels. Another example comes from Africa where the number of would-be viewers of mountain gorillas in Rwanda has been limited in the interest of their preservation. In cases like these, developers and managers, presumably, see the handwriting on the wall and move in the direction of sustainability. The real question, however, is when will they see that handwriting and act on it; and if they do act, will it be enough to promote sustainability? Consider a mental experiment. Suppose that the sustainable side gets key people in some tourism project to agree (whether out of fear, guilt, shame or some more positive sentiment) to go along with some kind of sustainable project. There still remain bureaucratic wickets to negotiate, each manned by functionaries bent on protecting their own turf. And on the state level, the project, while perhaps thought of as admirable in every way for the benefit of the local population, may be seen as conflicting with other state interests. For example, the need for jobs might prevail over a limit on construction in a particular resort. Here, the sectoral fragmentation, which often is to be found in public administration, could exacerbate conflict over the project.

On the other hand, de Kadt (1992:70–75), though not underestimating the difficulty of developing a policy of sustainability for tourism, sees a range of opportunities for the state to play a crucial role in sustainable projects. Some American environmentalists who have become disillusioned with working through government to attain their goals might dispute this and argue for direct action against, or with developmentalists, but it would seem foolish for them to overlook the sometimes difficult path towards governmental assistance in accomplishing their aims.

What view should an anthropologist working for sustainable tourism development take of all this? First of all, there should be an understanding that any tourism project involves transactions between a number of interested parties, often with conflicting agendas. Second, these parties should be seen as having varying degrees of power in establishing policy and seeing that it is implemented. The first order of business of the applied anthropologist working in this area should be to make an assessment of the power of the parties involved. Then, the identification of effective coalitions for accomplishing a sustainable course of action can be attempted. This course of action would seem to be more realistic than that suggested by the *Guide* of the World Tourism Organization which seems to argue that all involved in tourism development can somehow get together and benefit.

Ecotourism as an Alternative

A number of less developed countries – Costa Rica and Dominica being prominent among them – have begun to stress ecotourism in their development plans. Conceived in its narrower environmental sense, this is a more sustainable form of development, which aims at giving tourists an experience with nature (birds, animals, underwater life, scenery, etc.) that has a low impact on the environment. As Goldfarb (1989:7–11) indicates, this kind of tourism is mostly international and usually involves better-educated, urban European and North American nationals traveling to the rural tropics. Though it is growing rapidly, ecotourism still is not very important in terms of departing numbers (for the U.S., an estimated 2–3% of departing international tourists are ecotourists (Goldfarb 1992:8). However, it can be important in numbers of tourists for particular destination areas such as Kenya, which is so popular for viewing wild animals that traffic in game preserves is becoming a problem.

Because most ecotourism takes place in state-owned parks or reserves, the government often is heavily involved not only as an owner, but also enforcer (some privately developed eco-resorts have begun to appear, in Costa Rica, for example, but their share of ecotourism still is comparatively small). For this reason, Goldfarb makes the role of the host government pivotal in her analysis of ecotourism development. With its heavy involvement in ecotourism, the state has had greater influence than the destination area over what course to take. For example, the need for the development and maintenance of a national park will supersede a local need for subsistence. Thus, because of state-sponsored tourism development, old hunting patterns may be transformed into poaching; and as was pointed out for the U.S. Virgin Island of St John in *chapter 2* (see Olwig 1980), new state regulations for preserving the environment may prevent the local population from continuing old patterns of subsistence and force them either to leave or remain as tourism employees.

In these and other ways, according to Goldfarb (1989:35–37), locals tend to bear the brunt of problems associated with ecotourism development while, at the same time, most of the benefits are siphoned away to the state and those allied with it. This means that if benefits are to flow to the people who are most affected, it is especially important for the sustainable tenet of local participation to be followed. It goes without saying, of course, that pressures to increase the volume of tourist traffic beyond the carrying capacity of particular destinations have to be resisted. Goldfarb (1989:44–45) believes that international conservation organizations can lend a hand by supporting those projects which seem feasible and rejecting those which are not.

What, then, can be said in summary about the possible role of ecotourism in promoting sustainable tourism development? It certainly provides no panacea. As with tourism, generally, for some countries it simply is not a viable option, although a broadening of the range of attractions that may be conceived as points of ecotouristic interest (say, to include human heritage sites) might make it more viable than previously supposed. As Urry (1990) has stressed throughout his book, *The Tourist Gaze*, almost anything can be turned into a tourist attraction. A more limited version of this notion may apply to ecotourism.

Other countries, such as Kenya with its wild animals and Nepal with its mountains, are more fortunate in their natural attractions. They have rich touristic resources that attract ecotourists and can charge premium prices for visiting them. To maintain the viability of such resources, however, it is important that they stay within the carrying capacity of their various sites and not succumb to the ever-present inclination to increase the size of their touristic 'golden eggs' in ways that will ruin their attractiveness.

Finally, according to Goldfarb, there are the majority of destination areas, which have some, but not overwhelming attractiveness for ecotourists. For these countries ecotourism can be no more than an add-on option. Around the Cancun area on Mexico's Gulf Coast, for example,

there are a number of such sites; and if one wants to include arche-
ological or cultural points of interest as ecotourism destinations, the
"add-on" here can be considerable. Seen from an economic point of
view, ecotourism in such sites can be no more than a sideline to the more
serious business of mass tourism. Advocates of sustainable tourism can
only hope that, with the benefit of economies of scale, those involved in
the development and maintenance of mass tourism will be able to attend
to at least some of the tenets of sustainability.

The Bali Sustainable Development Project

Once again, to bring the discussion down from the realm of more airy
abstractions, consider an actual applied project, which, in this case,
involves sustainable tourism development on the island of Bali, where
overdevelopment of tourism, particularly in the south of the island, has
begun to take its toll. This is a strategic planning study of the Bali
Sustainable Development Project (BSDP), which, according to its (at this
writing) latest report, *Sustainable Development Strategy for Bali* (BSDP 1992),
has the intent of identifying the key Balinese developmental issues and
making recommendations to governmental agencies and other institutions about how to deal with them in a sustainable way. These recommendations were oriented towards possible incorporation in the latest
five-year development plan for 1994 to 1999.

Tourism, which has been the fastest growing sector of the Balinese
economy, with a projected doubling of international tourist arrivals to
about 700,000 for the five-year period ending in 1993, has begun to
generate some serious problems which also have been looked into by the
United Nations Development Programme (see Hassall *et al.* 1992). It is
seen by the Bali Sustainable Development Project as a particularly
important aspect of Balinese development, which must be understood in
terms of the whole. The studies, on which the report and other publications from the project are based, were carried out by faculty and graduate
students at Gadjah Mada University in Java, Udayana University in Bali,
and the University of Waterloo in Ontario, Canada. This is part of a
broader program of investigation funded by the Canadian International
Development Agency and administered by the University Consortium on
the Environment. A key figure in this project is Geoffrey Wall, whose
projections of effects of global warming on tourism were considered at
the end of the last chapter.

The BSDP report, which attempts to incorporate both 'bottom up' and
'top down' perspectives, makes use of studies conducted in a sample of
eight villages (about 70% of the Balinese people live in villages), such as
Bongkasa (BSDP 1991b) and Kelurahan Kerobokan (BSDP 1991a), and

extensive contacts with provincial agencies. Besides studies of particular villages, which have focussed on developmental stresses and the capabilities of the villagers to deal with them, there have been investigations which center on themes such as waste management and human resource development, sectors of development such as tourism and agriculture, and particular geographic areas such as coastal zones, forests and steeply sloping areas.

Throughout the research, regular workshops and broader discussions were held in Indonesia and Canada. At a final workshop prior to the writing of the report, 75 participants discussed the first draft of the report, and suggestions from this review were incorporated in the strategy to be presented. The authors of this pathbreaking report are conscious of the importance of their project. They argue that because of the renown of Bali, it now has an opportunity to serve as a model for sustainable development. They believe that the fate of such development, not only in Bali, but also in the world beyond, depends a good deal on Indonesian follow-through.

The report opens by pointing out that Indonesia already has taken a number of steps towards sustainable development, but that there are important stresses which need to be addressed either by the formulation of new policies or the readjustment of old. Consider some of the more important stresses. The loss of agricultural land and the use of agrochemicals raise questions about whether the population can be sustained with current agricultural practices. More and more waste is being produced by a growing population, an increasing number of tourist-visitors, and the run-off of dyes used in the garment industry; but methods of waste management are proving to be more and more inadequate. In coastal zones where tourism is important, coral reefs have been destroyed, sedimentation has increased, coastlines have eroded, and fresh water resources are becoming depleted. Some commercial developments, which do not conform to development regulations, threaten historic or sacred sites which have become points of touristic interest. Also, because development has been concentrated in certain parts of the island, not all of the Balinese are equally involved in it and so do not share in its benefits.

The report acknowledges the work of the Brundtland Commission, discussed earlier, and adapts the Commission's discussion of sustainable development to the Balinese cultural context. Besides familiar criteria of sustainability such as maintaining ecological integrity and social justice or equity (with special attention given to gender equity), there is, in accordance with Balinese values, a particular emphasis put on cultural integrity, which is conceived to include a balanced continuity of growth within the framework of the Balinese cultural heritage. The authors believe that an emphasis on cultural tourism, rather than beach tourism will contribute significantly to such growth.

After reviewing and applauding Indonesian and Balinese policies and actions that have supported the kind of sustainable development advocated in the report, its authors deal with various developmental faults, which include efforts to expand local forests (as a substitute for imports) that will produce wood for the handicraft industry (a plan which, if followed through, will, according to the authors, add to the already considerable pressure on agricultural land), efforts to develop tourism and agriculture, which have not taken into account possible inadequacies of the local water supply, and a failure to address an already grave problem of waste management and control. Though population density is high, population growth has dropped considerably as a result of a family planning program, so this issue, which often is salient in less developed countries, is not so here. The authors do think, however, that the issues of migration (of off-islanders seeking work in Bali, for example) and amenities needed to sustain an adequate quality of life for the present population need to be investigated.

Sustainable Development Strategy for Bali identifies a number of processes for developmental change. These involve the protection and enhancement of culture, with the dual concerns of stability and orderly change, the encouragement of economic development, in which the issues of balance and equity are stressed, the protection of environmental integrity, which has so far, according to the authors, not received much attention, the achievement of cross-sectoral integration, particularly regarding development planning and environmental management, the balancing of top-down and bottom-up approaches, in which the views of both the community and the various governmental levels are attended to, the achievement of partnerships, in which the various interests in Balinese society are considered, and finally, the legitimation of the principle of sustainability by government, in which strong and clear signals on sustainability are given and implementation is enforced in its favor.

The report sees tourism, which the government wanted to increase at a rate of 15% per annum on the international side and 5% on the domestic (goals which were considered too ambitious by the BSDP), as a major factor in its developmental strategy for Bali. Here, the prevailing emphasis on 'cultural' tourism, which includes religious and artistic presentations, visits to certain sites, and craft production, should, according to the authors, be increased. In addition, ecotourism (visits to Bali's many natural attractions) and agrotourism (visits to agricultural villages) should be developed. Here, attention should be given to minimizing impacts and offering appropriate information through well-trained tour guides and brochures or maps.

Though the village studies show that the vast majority of Balinese in the villages continue to be positive about tourists and tourism (see BSDP 1990), the authors of the report apparently still feel that visitor conduct can be improved. They recommend the preparation of a code of conduct to be distributed to visitors on arrival. Departing visitors also would regularly be asked to provide information about themselves and their experiences in Bali so that the quality of visitor experience and satisfaction can be monitored.

Because of the current overcapacity of hotel rooms in the Badung district in the south, the report advocates a moratorium on hotel building there. Rather, remediation or improvements should be made in a more sustainable direction. New hotel building in other areas conforming to sustainable principles also would be promoted, thus creating a more balanced, equitable process of development. All of this would be carefully considered in terms of the potential attractiveness of destinations, as well as the resources available, and would be financed by increased departure and accommodation taxes paid by tourist-visitors.

In a separate section, the report points out how the recommended process of tourism development would contribute to the objectives of sustainability. First of all, tourism, with its economic fall-out, obviously is development oriented. Then, if it is spread around (outside of its current concentration on the southern coast and involving females in more desirable positions, for example), it will contribute to social justice or equity. Next, with its emphasis on 'cultural' tourism (including agrotourism), the recommended process of development would strengthen the Balinese cultural heritage. And finally, through the promotion of ecotourism and agrotourism and the incorporation of ecologically sustainable principles in, for example, the building and maintenance of infrastructure, it could contribute not only to a respect for, but also the protection of the natural environment.

This latest report of the Bali Sustainable Development Project and associated materials that have been made available to the author are impressive for both their scope and depth. Sustainable tourism development is conceived as anthropologists would conceive it, that is, as one aspect of a total developmental context. Also, refreshingly, the connection of the proposed developmental process with the general principles of sustainability are spelled out. Here, because of local concerns and those of the Canadian Agency for International Development, which has provided much of the financial support for the BSDP, special consideration is given to the requirements of Balinese culture and to the issue of women's equity; but the authors still feel that they have managed to keep all of their recommendations within the minimal guidelines of sustainability.

The BSDP report and associated publications all show an acute sensitivity to what is currently a most important anthropological concern, which is to give the point of view of the 'other' its due. As an example of this, consider the associated study of tourism employment in Bali by Cukier-Snow and Wall (1993). These authors stress how important it is that researchers not jump to conclusions about the quality of tourism jobs. As was pointed out in *chapter 2* of this book, a frequent criticism of tourism-generated employment in the less developed world is that most tourism jobs for locals are of a menial nature (gardeners, maids, waiters, and the like), with most high-level jobs going to foreigners. But Cukier-Snow and Wall point out that from the point of the locals, menial jobs often are considered to be desirable and an improvement over, say, working in the fields.

Another strength of the report is the attention paid to Balinese villages in which the majority of the Balinese live and their actual and potential relationship to developmental issues. From information gathered by researchers in the sample of villages over a period of several months, it was possible to have a ethnographic base for the construction of the 'bottom up' perspective that has been lacking in so many developmental scenarios. This author's feeling that still more time could have been spent in the villages is only a minor quibble that probably derives from an anthropological prejudice that more fieldwork always is desirable.

Though the authors of *Sustainable Development Strategy for Bali* have done a good job in clarifying the various goals of sustainable develop-ment in the Balinese developmental context, they have been somewhat less successful in showing how potentially conflicting courses of action can be put together. That they are aware of such problems is indicated by, for example, their discussion of their recommended promotion of agrotourism, in which they argue that visits to agricultural villages should be spread around in time and place so as to minimize impact, which is to say that, where carrying capacities are threatened, they have opted to follow the well-known strategy of dispersal. Here, they seem to be following the same principle involved in their recommendation that no further tourism development be allowed in the Badung area, but that it be promoted elsewhere on the island according to their principles of sustainability.

However, the potentially corrupting effect of a greater amount of mass tourism on Balinese culture is not given much emphasis in this report. The positive consequences of tourism, acting through employment, are considered, but the consequences of having large numbers of tourist-visi-tors around and resulting demonstration and other effects, are hardly noticed; and, though the authors are fully aware of the anthropological studies of Bali (see e.g., Noronha 1979; McKean 1989; Picard 1979, 1990), there is no mention of one of the more important issues concerning tourism's consequences in the current anthropological and sociological literature, namely commoditization, which is supposed to transform everything into some measure of economic exchange. If tourism-induced commoditization does occur in Bali, what are the possibilities of following the BSDP's sustainable criterion of cultural continuity with a past in which many-sided communal activities were more important than individualistic activities dominated by the facts of monetary exchange?

In a personal communication, Geoffrey Wall, who, himself, is well aware of this issue, argues that the Balinese are willing to abide by a reasonable increase in commoditization if it means "better jobs, more money, better opportunities for their children, etc." Moreover, he feels that "the culture is strong and that tourism has given it added value in some respects." This is not to say that Balinese culture is not changing, nor that it will not change more through tourism and other development in the future, which is entirely in keeping with the vision of the future held by the BSDP. The principal threat of development is not, according to Wall, to the Balinese culture, but to the environment.

That tourism development is approaching or surpassing sustainable limits in this regard should be apparent to people visiting the Badung area where unrestrained development has taken place. One can now see for oneself the destruction of the coral reef; and when there no longer is sufficient water for Western tourists to swim, to bathe, and to drink, and inadequate management of human waste, even the enhancement of cultural tourism may not be enough to maintain the attractiveness of Bali as a tourist destination.

As mentioned earlier, the authors of the BSDP report are only proposing a strategy for sustainable development. This strategy is a forerunner for policy, which has to be worked out by government and private powers. How influential the report is in the development of policy and in its subsequent implementation remains to be seen. The people involved in the BSDP took extraordinary pains to work through the existing system and leave such matters to political and administrative people, as well as interested businessmen. They are heartened by the temporary ban on the construction of new hotels in the Badung district and the decision to entertain a UN-sponsored tourism plan (which, however, appears to have ignored the environmental issue).

Problems of implementation can be formidable indeed and depend ultimately on relevant structures of power. What are those structures in Bali? The authors of the report (BSDP 1992:27–30) recognize the problems of bureaucratic intransigence, lack of communication, a tendency to be moved primarily by friendship or kinship ties, faulty coordination, etc. They are also aware of the pulling and hauling between national interests, which tend to favor a strong push for tourism, and local interests, which (along with the BSDP) tend to favor more balanced development of tourism, agriculture and small industry. They recommend that the governor continue to rely on the "Expert Team of Bali Government", which consists of experts in a variety of fields of development planning, for advice on developmental matters.

But how sanguine can we be that the proposals of what is essentially a bunch of academics will be taken seriously by the political and business elite? The published materials do not give us much to go on, but based on what we know about countries such as Indonesia, one would expect that personal connections, not laid out on organization charts, and likely to be based on kinship and friendship ties, would ultimately prevail even over the interests of the Canadian Agency for International Development, which has been providing most of the funds for the Bali Sustainable Development Project. The recent death of a powerful local supporter of the BSDP's activities does not augur well for the future, but attempts are being made at this writing to develop other important 'patrons'.

Those involved with the Bali Sustainable Development Project are, naturally, hoping that everything works out in line with their proposals, but Wall, for one, seems to be resigned to accept less than that. He says (in a personal communication), "As outsiders, we can raise questions and put items on agendas, which may be difficult for local people to do". He and his associates can take some pride in having produced something important in the literature on development, and a number of these scholars are in the process of turning their experience in the BSDP to academic account. But one wonders whether this is enough for people who, every day in their work on Bali, are reminded of the pressing need for sustainable development there.

* * *

Before ending this chapter on sustainable tourism development, the role of tourist generating areas in the attainment of sustainable goals in destination areas ought to be considered. The authors of *Sustainable Development Strategy for Bali* bring up this matter tangentially in recommending the distribution of a code of conduct to arrivals who, presumably, have not had a course on Balinese culture and whose ways may conflict with the values of sustainability.

Even though these visitors may recycle, take public transportation, vote for environmentally friendly candidates, and insist on community participation in developmental projects at home and, where possible, abroad, most of them do live in an industrialized society, which, by its very nature, violates most of the tenets of sustainability, however it is defined. The culture they represent and 'carry' with them to places like Bali on their tourist odysseys is bound to contravene to some extent whatever host principles for sustainable development are in place. A particularly dramatic example of this is the need of Westerners for water – to bathe in, consume, keep cool, etc., which inevitably puts strains on certain aspects of the tourist infrastructure.

Here, one is reminded once again of the prescient analyses of Krippendorf (1986, 1987), who points out that recreation and tourism are integral parts of the whole industrial social system. In such societies, according to him (Krippendorf, 1986:520), the environment "is treated and exploited as if resources were inexhaustible and infinite". Though he also (Krippendorf, 1986:525) conceives of tourism as a kind of social therapy, it is therapy in the service of industrial values, which means that tenets of sustainability, while touring, can only be tolerated in comparatively small doses, as, for example, taking an afternoon walk in the forest.

What ultimately is necessary for sustainable principles to prevail among industrialized tourists, according to Krippendorf, is a complete reorganization of industrial society itself. Certainly, one can see evidence of the advance of sustainable values – compare, for example, the atmosphere of Pittsburgh, Pennsylvania today with the way it was twenty-five years ago – but whether one can be as optimistic as Krippendorf (1986:525–531) about the coming "profound change", in which the load limits of ecosystems are recognized and fully accepted in the advanced industrial nations, is an open question. Also, can one expect a move by societies such as China, which have only recently embarked on a process of forced industrialization, to constrain their development in the interests of sustainability? For them, that might be a luxury they do not believe they can now afford.

Finally, it should be pointed out how research on sustainable development can fit into the anthropological agenda. Like all applied projects, sustainable development involves some practical value that is supposed to be realized. An applied anthropologist may or may not subscribe to that practical value, but by coming on board, he or she is working for its realization. No significant problem here. Sustainable development, in one form or another, is something that most anthropologists would support willingly. Then, their first order of business should be to clarify the term sustainable development as it applies to their project. How is this specific kind of sustainable development to be promoted in this case? Is it feasible in this situation? What are likely to be the consequences of the courses of action being proposed? Are there other acceptable courses of action? All of this requires that the relevant touristic field of action must be laid out, with the interests and the power of the parties specified. Only after this has been accomplished, can the possibilities of the success of the particular project be gauged. The extended discussion of the Bali Sustainable Development Project, included in this chapter, should have provided a hands-on feel for some of the problems involved.

Chapter 8

Basic and Applied Research: Hand in Hand?

What do Disputes About the Nature of the World Tell Us?

In the history of science, there is no end of disputes about what the world is like. What is ostensibly the same phenomenon looks different to different scientists. Such disputes help us to improve our comprehension of the nature of the world, but also tell us something about the researchers involved and the procedures they employ. This chapter will be concerned not only with the usual factors affecting research, but also additional factors associated with applied research. In the anthropological study of tourism, what are the special problems posed by applied research? Can they be resolved for the benefit of more basic concerns and for anthropology as a whole? The issues involved can be introduced by exploring disputes about the nature of human activities, including tourism.

In tourism research, one well-known dispute, which has been mentioned earlier, concerns the intentions of the (modern) tourist. What are these tourists really up to? Boorstin (1964) saw these tourists drawn towards inauthentic "pseudo-events", which is a spillover from a superficial life at home. MacCannell (1976), on the other hand, saw the tourist as rejecting that superficial, inauthentic home life in search of more authentic experience abroad as a compensation for life at home. Which of these authors is correct? By asking and attempting to answer this question, Cohen (1988a, 1988b), for one, helps us to get a better grip on the exact nature of the reality of tourist inclinations, and so to advance the scientific study of tourism.

One reason for MacCannell's dispute with Boorstin could have been that the two authors were observing different phenomena or different aspects of the same phenomenon. In anthropology, this has been one of the reasons given for the famous discrepancy between the views of Mead (1949) and Freeman (1983) concerning the character of the people of American Samoa (see e.g., Levy 1983, 1984, for an analysis of the dispute), who, according to one Samoan quoted by Levy (1984:87), were taken to be either "free-loving orgiasts" (a caricature of Mead's view) or "sex-starved rapists" (a caricature of Freeman's). That population was variable, it was said, and these authors had focussed on different sub-populations. Mead, for example, concentrated on adolescents in an area, which, according to Levy (1983:831), "may have had features that made it, in the 1920s (and later), different from other parts of Samoa". In regard to the Boorstin–MacCannell dispute on authenticity-seeking tourists, Erik Cohen also resorts to the variability of populations to explain it. Though he believes that all modern tourists are concerned with having some minimum of authenticity in their experience, some are looking for it more than others. He (Cohen 1988b:376) says, "It follows that intellectuals and more alienated individuals will engage in a more serious quest for authenticity than most rank-and-file members of society".

Another possible reason for the difference between views of some phenomenon could be the research procedures involved. Levy (1983:831) feels that neither Mead nor Freeman adequately "contextualize" their materials so that they can be properly evaluated as data. Further, he (Levy 1984:89) believes that they both were selective in their data gathering, with Mead tending to rely on the rhetoric of adolescents while Freeman relied on that of "the religious and political guardians of order". Neither MacCannell nor Boorstin are strong empiricists. MacCannell (1976:4) seems rather casual. He says that he collected his data by "following tourists, sometimes joining their groups, sometimes watching them from afar through writing by, for and about them". Boorstin, who made no claim to being a scientist, was still less concerned with empirical details and freely mixes opinion with his facts in making his point. Though one gets the impression that MacCannell has the upper hand empirically, there remains a nagging feeling that, ultimately for both of them, facts will give way to opinion. Cohen (1988a:41), while recognizing the importance of the views of these men for the advancement of scientific research on tourism, seems to be fully aware of their empirical failings.

Another factor, theory or ideology, could have made a difference between the two men's views. Boorstin and MacCannell do, indeed, appear to have had different theoretical or ideological axes to grind. Boorstin was engaged in tracing an historical transition in the West from travelers (supposedly looking for real experiences) to tourists (supposedly satisfied with inauthentic pseudo-events). This preoccupation, which is not, in fact, a scientific theory, but rather a widely shared prejudice against modern tourists, could have led to an overstatement of the differences between the two kinds of travelers. For him, the modern tourist has few redeeming virtues. MacCannell, though he seems to be equally critical of modern society, is not so critical of its tourists who are seen as trying to escape from a frustratingly alienating modern existence.

Obviously, this view owes much to theories propounded by people like Marx, Durkheim and Max Weber who were critical of the society in which they lived.

As far as Freeman and Mead are concerned, the theoretical underpinnings of their views on Samoan character are clear. Mead was following the theoretical program of the pioneering German-American anthropologist, Franz Boas, which emphasized the importance of culture (socially learned experience) in the formation of character, while Freeman, though not denying culture as a factor, wants to bring biology back into the picture. Thus, his views of people troubled by sexual conflict reflect his belief in the biologically and socially dictated inevitability of such conflict, while Mead, following the notion that such conflict will vary according to the pressure of social circumstances, believes that in Samoa, a benign socializing experience has reduced sexual conflict to a minimum (see Levy 1983:830, for an elaboration of this).

So much for the strictly scientific reasons that might account for the discrepancy in views concerning modern tourists, offered by Dean MacCannell and Daniel Boorstin, and Samoans by Margaret Mead and Derek Freeman. They relate to conclusions about some aspect of the world 'out there' that these people chose to investigate, the procedures they used in their investigations (methodology), and the particular perspective from which the observations were carried out (theory). Kitcher (1993:164) refers to these factors as involving input from nature (in this case, sociocultural reality) and the practices of a scientific community, which certainly are involved in the constructions of these authors.

We might want to add here some reference to the process of scientific construction, especially in sociocultural anthropology, throughout which the factors being discussed here are operative. Regarding Mead and Freeman, Levy (1984:87) says that "they did their first construction in the course of their fieldwork, and then, in a second creative movement, from the oral and experiential to the written mode". Other constructions earlier and later in the process could be added.

There are still additional reasons, not strictly scientific, for drawing different conclusions about what, ostensibly, is the same phenomenon. They include the personality of the scientist and his or her social or political position. Consider first the personalities of the four authors and their life histories. Could there be something in those histories – in the families in which they were raised or their later experiences, including fieldwork – that might have biased their investigations? With or without their cooperation, we might be able to build a case concerning such influences. As has been indicated earlier in this book, anthropologists have been increasingly willing to bring this kind of analysis into their work, and there is a growing literature on the subject. Nash and Wintrob

(1972), who traced and attempted to account for the emergence of self-consciousness in ethnography, were among the first to recognize and discuss this subjective trend in anthropology.

What about the political or social positions of the authors? Could they have had a biasing effect on their work? We are informed best about the position of MacCannell, who, in *Empty Meeting Grounds* (1992:1–13), gives us an extended disquisition on both his theoretical and political position. As was pointed out in *chapter 2*, he does this so that the reader can take into account his liberal–radical stance in evaluating his account of touristic encounters. His picture of socially alienated tourists trying to make up for that alienation abroad undoubtedly is informed by this stance. MacCannell's is one kind of self-reflection that has become increasingly acceptable in anthropological work.

Perhaps one doesn't need to have such a declaration from those authors working in the Marxist theoretical tradition, for whom theory and praxis often go hand in hand. With them, a leftist political position almost certainly implies some kind of class analysis organized around exploiters and the exploited, a view that informs a good deal of the criticism of tourism development in the Third World.

All of the factors just mentioned can account for differences in the ways in which scientists go about asking and answering questions about reality, and so, ought to be included in any critical analysis of a scientific project. One additional factor needs to be considered. This is the applied factor. When the questions a scientist puts to the world are not dictated strictly by scientific concerns, but by practical matters, he or she may be thought to have put on an applied hat. For example, scientists who do consulting research for the World Tourism Organization (WTO) are routinely asked to come up with recommendations to host governments concerning tourism development. Thus, addressing the problem of (until then) meager benefits from tourism in the Seychelles Islands, consulting investigators of the WTO recommended that charter flights to the islands be terminated, a restructuring of agreements concerning hotel ownership instituted, the total tourist carrying capacity of the islands somewhat increased and a program designed to attract more up-scale tourists, particularly from the Pacific rim, be inaugurated (unpublished WTO analysis). Neither MacCannell nor Boorstin were much engaged in such work, but the applied side of the anthropologically oriented study of tourism certainly cannot be ignored in evaluating anthropologists' findings.

If we agree that the aim of anthropological study of tourism is to contribute to our understanding of the touristic process, and ultimately of the human condition, there is no *prima facie* reason why applied tourism studies cannot add to that contribution. There are special problems associated with this kind of investigation, but they are not of a different order and can be dealt with if researchers are aware of the problems involved and act in terms of that awareness. In addition, there are opportunities in such work not only to advance the study of tourism, but also the entire anthropological enterprise.

Opportunities and Hazards in Applied Tourism Research

Let us agree, then, that no scientist is totally free to apprehend the world as it is. The picture that scientists, including anthropologists, construct of the world is affected by a number of factors including the procedures and theories they employ, what kind of people they are and the social context in which they work. The constructions of applied researchers dealing with more practical questions are affected not only by such factors, but also by the control of clients – real or imagined – over their work. These clients have more or less say over what questions will be raised, how they will be addressed, and so, how the work will come out. As a result, applied investigations take on a somewhat different character than those of basic science.

In an interesting comparison, Butler (1992:138–139) lays out what he believes to be the general characteristics of three kinds of tourism impact assessment carried out by social scientists: that conducted by academics (who, presumably, are more likely to be practicing basic science), research for private clients and that for clients in the public sphere. He believes that though there are a good many exceptions, these three kinds of investigations tend to differ in a number of ways. Butler (1992:138) points out, for example that "private sector studies... have traditionally been predictive studies, generally concentrating on economic impacts... and focussing on the benefits of tourism to proponents and communities", while studies done by academics "for the most part, have been post-development assessments and have often focussed upon the negative or undesired effects of tourism and related developments". As far as the entire course of the anthropological study of tourism is concerned, this seems to be a fair appraisal, but one should note that there has been an increasingly neutral tone in both basic and applied anthropological work on tourism's impact on host societies and their environment. Ritchie's (1992:206) suggestion that academics, in contrast to applied researchers, "though also concerned with strategic issues of immediate concern, often have an interest in more fundamental or basic research addressing major issues with medium or long-term implications" also seems to ring true for the anthropological study of tourism.

As has been indicated throughout this book, anthropologists have been interested mostly in the host or destination end of the touristic process, in which a concern with tourism's social impact has predominated. This interest has been advanced not only by the natural concern of anthropologists with areas where many of them have traditionally done their work, but also by the interested support of public national and international agencies that have funded such work. Interestingly, there is some evidence that indicates that this kind of emphasis may not occur in tourism research in countries with centrally controlled economies such as China or what was once the Soviet Bloc.

Perhaps anthropologists in academia could do more to demonstrate the applicability of their more basic scientific views and procedures to government and private clients. They could show, for example the value of their comparative perspective in sketching out alternatives for some particular developmental scheme. They could stress the usefulness of their habit of getting to know the 'other' in gaining acceptance of some developmental plan by a host people. They could show that they are knowledgeable about, and interested in subjects of interest to clients. They could make an effort to talk the same language – something that they do routinely with subjects in the field. And finally, they could make an effort to sell themselves and their work to public and private clients, not wait for those clients to come to them.

The bottom line, however, is that to date, for whatever reasons, there hasn't been a great deal of demand for applied research on tourism by anthropologists, and that what demand there is concerns almost entirely destination areas. Valene Smith (1992:3), who has been an indefatigable promoter of this kind of research, says that "career opportunities abound, especially for individuals who 'create' their jobs, as anthropologists have always done when working beyond the University". So far, however, the growth of full-time jobs for anthropologists in applied work has not matched the dramatic growth in tourism around the world which she outlines. Perhaps opportunities will increase with a continued expansion of tourism in the future, but what has happened so far in the relatively brief history of the anthropological study of tourism would seem to indicate that opportunities for this kind of work will not be easy to come by.

In applied projects the influence of clients on the research is most apparent in the nature of the questions asked. Consider some examples of applied anthropological work (in destination areas only) which obviously are associated with interests of clients. A private foundation, the United States Government, and a state university provided funds for research designed to answer questions about how to attract new tourism to small towns in northeastern California which were experiencing the effects of a downturn in lumbering operations (see Smith *et al.* 1986). Evans' study (1986) of factors limiting Indian tourism development in California is organized around questions of interest not only to local Indians, but also the California State Department of Tourism and the

State Department of Parks and Recreation in which she was employed. Jagusiewicz' (1990) study of the limitations and prospects of tourism development in Poland's Great Mazurian Lakes region, prompted by the 'progressing ecological threat' of tourism development, was carried out under the aegis of (then socialist) Poland's Institute of Tourism. And an environmental assessment for the Eastern Caribbean by Edward and Judith Towle (Towle and Towle 1991), in which tourism-implicated questions of progressive environmental degradation were raised, was carried out under the jurisdiction of the Caribbean Conservation Association and the Island Resources Foundation with financial aid from the U.S. Agency for International Development.

All scientists face hazards associated with their scientific missions. In *chapter 1*, some of the problems that are peculiar to anthropology (e.g., getting to know the 'other' and culture shock) were discussed. The special problems of applied anthropology involve the practical interests of clients which conflict with formal or informal anthropological guidelines. Certain of these guidelines involve all anthropologists, more or less, and so have the status of cultural norms of the discipline as a whole. Others involve sub-groups within the discipline, and so, are sub-cultural in nature. It probably would be best for anthropologists to assume that some kind of conflict over such norms will emerge in the course of their applied work.

First, consider an hypothetical example, which could have occurred, but actually did not. In *chapter 6*, an unpublished plan (perused by the author) for the tourism development of a little French dependency in the Caribbean was discussed. This plan was based on an investigation of the island, its resources, the potential market for tourism there, etc. It was done by a French consulting firm. No anthropologist appears to have been involved in the investigation, upon which a proposal for the tourism development of the island was based. That proposal, it will be recalled, failed to take into account the island people who would be directly affected by the tourism development. Nor were these host people significantly involved in the formulation of the proposal, which suggested that three areas of the island be used for the development of three different kinds of tourism.

If anthropologists had been involved in this project, the failure to include hosts either as subjects or participants in the creation of this developmental scheme should have been intolerable to them. Anthropologists these days normally expect that any people likely to be affected by some social project should participate in working it out; and the *Principles of Professional Responsibility* of the largest association of anthropologists, the American Anthropological Association, state that, for the anthropologist, the interests of subjects and science should be foremost. In the hypothetical case, those host people who would be centrally involved in this developmental scheme were being largely ignored. How would our hypothetical applied anthropologist have adapted in a situation such as that?

Howell (1994) offers a real example of problems facing an anthropologist working as a cultural specialist in heritage tourism projects. Speaking from her vantage point of experience as director, consultant and evaluator for several community history projects supported by the Tennessee Humanities Council, she lays out the problems of an applied anthropologist in working through these and other heritage tourism projects.

Howell doesn't tell us too much about her own views on what form such projects should take, but a normal anthropologist, with an antipathy for elitist views of history, probably would have had some populist view or other. Whatever Howell's views were, they had to contend, during the course of the projects, with those of developers with an interest in making tourist attractions popular and of various community groups, each with a stake in some version of a community's history. Further, the cultural specialist had to deal with the special interests of a project director and various functionaries from the local level on up to the United States Government. Finally, he or she had to deal with the critique of cultural conservation and heritage tourism by some anthropologists. That critique, put forth most forcefully, perhaps, by Handler (1988, 1994), whose work on the bureaucratic production of history leads to the conclusion that local peoples involved in such projects are victims of bureaucratic machinations, and Greenwood (1977), who was among the first anthropologists to raise the issue of tourism-induced commoditization, was taken seriously by Howell.

In Howell's case, there are a variety of clients, each of whom has an interest in how a project should develop. There are also some anthropologists who are more or less critical of the value of such projects. Not mentioned, but probably internalized by the anthropologist working as cultural specialist, are the requirements for normal anthropological work which have been mentioned above. And one should not forget about the personal values of the anthropologist herself in toting up the numerous conflicts facing the cultural specialist in such projects. Considering all these more or less incompatible interests with which the cultural specialist had to contend in helping to develop an appropriate historic attraction, it is no wonder that Howell (1994:157) adopts Maya Angelou's view that "courage is the most important virtue".

Is it possible to find a case where the applied anthropologist actually is faced with client pressures to bend the truth in the course of investigating and making recommendations about some tourism phenomenon or other? One thinks back to the summary case study of Mansperger's work on Yapese tourism development that appeared toward the end of *chapter 6*. Mansperger, it will be recalled, had come to the conclusion that one of the biggest problems facing the Yapese was a severe imbalance of trade. If the Yapese would import less, they could possibly adopt less heroic developmental measures, including touristic, to satisfy their needs. Yet nowhere in his recommendations for tourism development on Yap does Mansperger allude to the issue of foreign imports. Perhaps he had concluded from his conversations with Yap officials that mentioning this issue would serve no purpose because it was beyond the range of possibility. Perhaps he had concluded that it would be better not to mention the fact that tourism development, itself, often is associated with increased imports. No matter. The effect is the same. In this case, the whole truth, no matter how unpalatable, is not mentioned, possibly because it was not considered compatible with a client's interest.

Other conflicts resulting from the incompatibility of client and anthropological interests in tourism research might be mentioned. Certainly, the requirements of the new reflexive anthropology demand that the applied anthropologist be aware of them and deal with them. In the extreme case, that anthropologist, confronted with what he or she construes to be the "selling out" of some anthropological interest or other, might decide to withdraw from a project. In many cases, however, some rapprochement between anthropological and client interests can be worked out. In doing this, the anthropologist should be aware that it is not only an issue of damage control where applied work is concerned. There may be an opportunity to actually improve the lot of the people involved and to further (through knowledge gained or conveyed) the interests of the discipline.

Howell's analysis makes clear what some of these benefits to anthropology might be. As pointed out in *chapter 4*, one of the enduring interests in sociology and anthropology has been in how basal social structures influence social superstructures such as tourism. In the course of the creation of an historic attraction, during which various contending groups pushed their own versions of their community's history, Howell was able to witness this process in action. She (Howell 1994:156) says, "But our ethnographic encounters in the host community soon begin to suggest the meaning different segments of the community attach to their history, how these meanings are played out in present-day social interaction, and how these social dynamics inform differing representations of the past". Here, the applied anthropologist has an opportunity to do something that probably is not stated in the formal contract, namely, acquire theory-relevant data which could be of use in some more basic research.

Howell also argues that in their consulting roles anthropologists may have an opportunity to influence the world of affairs. In the course of consulting for some historic project, for example, the cultural specialist has an opportunity to bring an ethnographic perspective to the table, and so, communicate an anthropological viewpoint to others. She (Howell

1994:155) points out that "such consulting opportunities are welcome means of supporting cultural diversity and communicating regional experience which resonates with recent developments in the post-industrial nation at large...". Those anthropologists who are passionate about sharing their anthropological vision with students and others outside the discipline should find those applied projects, in which they have an opportunity to advance their anthropological view of things, particularly attractive.

Sometimes, scientists have blind spots which are the result of a lack of access to certain aspects of the world. We have seen that, as far as anthropological research on tourism is concerned, one such blind spot concerns the tourist and tourism generating situation where the root causes of touristic phenomena are at work. There, people with higher powers initiate and control tourism development, which affects others in destination areas, including anthropologists who may be studying them.

Now what if an applied anthropologist could gain a position as consultant in such a project? This would make him or her privy to the deliberations of the powerful, which could lead to a direct understanding of the ultimate causal processes involved in the project. Because of this involvement, therefore, the applied anthropologist would be in a privileged position to illuminate, through their own and others' work, what may formerly have been an inaccessible aspect of the touristic process and thus to add to our understanding of sociocultural life.

In this section, the problems and opportunities of applied anthropological work on tourism have been discussed and examples given. The argument has been advanced that problems of such research all derive from some kind of incompatibility between client and anthropological interests. A number of examples were given to illustrate both problems and opportunities. If applied anthropologists are aware of whatever incompatibilities exist between their practical and scientific missions and deal with them constructively, it may be possible to take advantage of the opportunities which such work provides for the advancement of anthropological understanding.

Hand in Hand

In their article on anthropology and tourism, Nash and Smith (1991:20) say, "The important point is that both theoretical and practical concerns can be addressed at the same time and that research towards practical ends can pay theoretical dividends". This can happen when one person plays both basic and applied roles or when two or more scientists share their work. It would be best, perhaps, to keep in mind that the flow of information is not necessarily one-way, as when applied data are used for theoretical purposes either by the same person or others, but can involve a two-way give and take. There follow a number of examples that illustrate this process in action.

Package Tourism

Earlier in the book, reference was made to the recommendations of the consultants of the World Tourism Organization to the government of the Seychelles Islands which had become dissatisfied with its income from an expanding tourism. Among those recommendations were those that suggested that charter tourist flights to the Seychelles be terminated and that the hotel industry on the islands be restructured. We are not privy to the investigation of the consultants which provided the basis for these and other recommendations. Possibly the consultants found out from their own investigation in the Seychelles that, because much of the tourism involved pre-paid packages, including travel and accommodation in hotels owned by the charter companies, its benefits to the Seychelles were minimal (cruise tourism is another example where benefits to the hosts are minimal). Or perhaps they were aware of earlier applied studies of the subject by others (see e.g., Estrada 1973; Spill 1976; Wahab 1975). In any case, the consultants were acting in terms of what was becoming a well-known fact in the tourism industry.

As Douglas Pearce (1987) makes clear in his study of package tourism in Europe, the tendency in the industry has been towards greater vertical integration (as when charter airlines are taken over by travel agencies) and horizontal integration (as when charter airlines become the proprietors of hotels in destination areas). That tendency, which is a fact of life deriving from the contemporary tourism generating situation, tends to shape less powerful destination areas in ways that provide fertile ground for the application of dependency theory. Pearce does not miss the implications of his own investigation for the testing of this kind of theory. He refers to the work of Oglethorpe (1984) on Malta, where U.K. package tour operators dominate the local market and speaks of a crisis of dependence there. He could have used the work on the Seychelles of the consultants of the World Tourism Organization to the same effect.

The Role of Residents in Tourism Development

Though possibly limited in its cross-cultural applicability, the notion that a positive attitude by local people will assist in any development project, including tourism, has become almost a truism (but recall Pearce's [1992,:26] critical observation about hosts with developmental interests mentioned earlier). To begin his applied investigation of predicting factors associated with attitudes towards tourism development among residents of two communities in the Columbia River Gorge in Oregon, Samuel Lankford (1992:225) offers as an assumption his own version of this taken-for-granted notion as follows: "Resident attitudes towards visitors and the accompanying use of facilities have a direct effect on the successful implementation of tourism development strategies". Lankford's study, which involved a telephone survey of local inhabitants, was carried out with the support of a considerable number of governmental and para-governmental organizations and agencies ranging from those associated with local municipalities through the state to the national level. The investigation not only responds to the interests of these various 'others', but builds on past research that associates attitudes towards tourism development among residents with socio-demographic variables such as length of residence, level of contact with tourists, and economic dependency on tourism.

A number of past statistical associations between such variables and attitude towards tourism development were confirmed by this research project. In addition, the use of multivariate statistical analysis helped to sort out the specific effects of different variables. Lankford demonstrates, for example the specific effect of reported direct, personal involvement in a developmental issue on a person's attitude towards tourism development. He (Lankford 1992:226) found that among those residents "who found that tourism was having an increasingly negative impact on their own outdoor recreational opportunities, the desire for further tourism development decreased".

Lankford and his research associates took great pains to keep all those with a direct interest in the study informed. A preliminary meeting was held to explain the importance of the research and its potential uses; everyone was kept abreast of the progress of the research; and a final presentation was offered, during which the implications of the research for policy and planning were discussed (personal communication). Once again, we are reminded that applied research, particularly if a number of interested parties are involved, can involve significant organizational and diplomatic problems.

As pointed out above, Lankford was aware of basic scientific implications of his study and was hoping to contribute to this aspect of tourism research. Indeed, his research stands out in a field of study, which, according to Ap (1990:614), is notable for the "atheoretical orientation of its studies". He mentions, for example that some authors (see e.g., the work of Butler [1980] mentioned in *chapter 2*) have developed evolutionary or cyclical schema for resort development. One of these schemes, which was developed by Doxey (1975) during research in Barbados and Niagara-on-the-Lake in Ontario, Canada, seems to him to be particularly relevant. Doxey's argument is that the attitudes of resort residents towards tourism development pass through successive stages ranging from euphoria to antagonism and suggests factors in the contact situation that are responsible. Certainly, Doxey's scheme cannot be taken as inevitable and should be applied with caution in other social contexts, but his concern with specific factors that generate irritation among residents fits in nicely with Lankford's applied work.

Finally, viewed in terms of the acculturation or development paradigms discussed at length in *chapters* 2 and 5, Lankford's study makes an important contribution. Too often, scholars using such paradigms omit the actors involved in the contact situation. Only cultures, which tend to be thought of as the ways in which groups of people tend to act, are seen to be involved in contact, not the people who create and reproduce them. It is important, always, to keep these actors or agents in the picture, as authors like Bidney (1967) and Giddens (1979) have reminded us. As far as tourism development is concerned, the attitudes of residents surely must be considered if one is to predict what is going to happen. One would hope, however, that somewhere in such research there will be a recognition of the power of individuals or groups of actors. After all, a negative attitude of residents towards a tourism development project should prove to be more or less obstructive according to the power they wield.

Tourist–Host Communication

Because tourism involves contact between cultures or subcultures, there are bound to be problems of communication between the tourists and those among the hosts who are directly involved in providing for their needs. Hotels and restaurants, for example can become aware of such problems because of what may turn out to be costly misunderstandings associated with menu orders, laundry services, accounting and the like. One way of dealing with such problems is to improve employee fluency in guest languages, and applied scientists have worked with multinational hotel chains to improve service workers' ability to communicate with their guests. An example of this is to be found in the development and evaluation of a training program, including training in cultural sensitivity and some essentials of the Japanese language, for employees of a Sheraton Hotel to communicate better with Japanese guests (see Sage 1989).

Often in situations of tourist–host contact a special system of communication emerges which is organized around the more commonplace needs of the tourist-visitor. It is this tourist language (TL), rather than a foreign language (FL), that may be the more important communicative skill; and it is on this language that formal courses for employees or prospective employees may concentrate. Speaking about bungalow tourism in the southern Thai islands, Erik Cohen and Robert Cooper (1986:550) report that all owners recognized the importance of fluency in TL in running their businesses.

Cohen and Cooper use the practical significance of TL as a take-off point for an exploratory sociolinguistic study, which focusses on the social context of languages in contact, in particular, that involving tourists and their hosts. Perhaps the most important defining characteristic of this contact is the fact that the tourist is a genuine stranger, who is, in Simmel's famous words, "here today and gone tomorrow". The language that develops in such a situation is one of a variety of simplified registers involving strangers, of which pidgin may be the best-known example. In their exploratory study, Cohen and Cooper discuss the nature of TL, the principal quality of which is its orientation towards the tourist's needs and demands. These authors point out that TL takes different forms according to the nature of the tourists involved. Using Cohen's well-known scheme for classifying tourists in terms of their adventuresomeness, they distinguish between institutionalized mass tourism, in which a highly standardized TL is used, and non-institutionalized tourism off the beaten path in which the TL is less standardized and more accommodating to the hosts. They go on to raise a number of sociolinguistic questions about tourist–host contacts and associated TL's for further investigation, all of which can easily be fitted into the acculturation or development paradigm that has informed so much anthropological work on tourism.

The Promotion of Sustainable Tourism

One method that has been used to promote sustainable tourism (sometimes called, in the jargon of the trade, a threatened environmental management device) is to orient tourists beforehand with some written information (lectures and discussion also may be used). This method was included, it will be recalled, in the recommendations of the Bali Sustainable Development Project (reported in *chapter 7*) and those of Mark Mansperger to the government of Yap (reported in *chapter 6*). The American Museum of Natural History, which has been conducting natural history tours since 1953 and was growing concerned about the tourism-induced degradation of the natural habitats of the world, wanted to explore the use of such a device by organizations having a stake in sustainable tourism development. This would be a first step in developing a set of ecotourism guidelines that could be made available not only to its own clients, but others' as well. Accordingly, it put researchers to work to collect guidelines that were being used by environmental organizations, tour operators, public land managers, outdoor equipment stores, and religious organizations. The project and its findings are reported in the research note, *Ecotourism and Minimum Impact Policy* (Blangy and Nielsen 1993).

Not all respondents to the researchers' letter of inquiry reported that they had guidelines of their own. Some said that they relied on destination areas for the orientation of their clients. Those 60 guidelines on which the report is based had a number of goals related to sustainability such as minimizing visitor impact and promoting conservation projects. The literature was aimed at a variety of audiences. From the authors' description, one gets the impression that it could not be characterized as employing sophisticated marketing techniques.

How to improve the guidelines? The authors offer a number of recommendations, which include the application of the latest (Western?) techniques for efficient communication, the development of both general and site-specific guidelines and better collaboration within the tourism industry. One cannot help but feel that, inasmuch as they offer no confirmation of the effectiveness of their proposed courses of action, they are somewhat ambitious in this regard. They probably are right about one thing, however: orienting guidelines can make only a very small contribution to sustainability.

This exploratory research note does not offer very much in the way of analysis of content. Such an analysis would provide significant information about what the creators of the guidelines were up to. Earlier in this book (recall, especially, *chapter 4*), a number of symbolic analyses of this sort were mentioned, as for example, of Japanese travel brochures or picture postcards of American Indians of the southwest. Here, additional anthropologically oriented analyses by Riemer (1990) and Selwyn (1990) of images produced by Canadian and British tour packaging firms, which are in the business of creating what Riemer refers to as "meaningful dream vacations", should be considered. Both of these authors see the travel literature of these firms as reflecting and anticipating the preoccupations of people in their societies. For example, Selwyn's apparently preferred 'reading' of the content of his British travel brochures, stresses the growing commoditization of the contemporary world, which is one of the more important current theoretical themes in anthropological and sociological research on tourism. This theme arises from a view of contemporary postmodern or postindustrial society as consumption, rather than production oriented, in which symbolic manipulations of the public have come to the fore.

It may not be so easy to think that the writers of guidelines for sustainable development and those creating brochures for packaged tours are in the same 'business'. Certainly, one group appears to be more businesslike than the other. But both do appear to have the same kind of goal, which is to persuade clients to pursue (consume?) certain kinds of tourism. In doing this, they must attend to, and work with people in the touristic process who are directly connected with their work. Applied analysis, with assistance from market segmentation research, can help them manipulate these people. It also can contribute to our understanding of how tourism is generated in a society.

The Effects of a Touristic Alternative

The search for benign alternatives to mass tourism continues. One such alternative, ecotourism, was analyzed in detail in *chapter 7*. Another involves some kind of integration of tourists and hosts, such as the famous *tourisme rural integré*, the brainchild of Christian Saglio (1979), which has been developed in the Lower Casamance region of southern Senegal near Ziguinchor . This kind of tourism, which now is 'on hold' because of secessionist struggles in the area, involves small numbers of tourists who come to live in 'traditional' villages in dwellings having only basic Western amenities such as toilets and showers. The local population has been fully involved in the planning and operation of these projects, and most of the benefits accrue to them. The tourist-visitors, who are mostly younger, more adventurous types, live next door to local people and participate in their activities.

Does this form of tourism, in fact, do what it is supposed to do, that is, minimize the deleterious consequences of mass tourism? Marguerite Schlechten, a doctoral candidate at the University of Fribourg, Suisse (Switzerland) set out to find the answer to this practical question. Her published dissertation, though not without defects, is something of a high water mark in research on tourism development in that it combines a full understanding of the subject of tourism with considerable sophistication in theory and methodology.

Schlechten had prepared herself for her doctoral effort with work on African tourism for her license. As indicated by the title of her thesis for this license, *Alienation ou rencontre* (1978), she seems to have been particularly interested in the nature of the relationship between tourists and their hosts, which is a well-established subject in tourism research. For her doctoral thesis, *Tourisme balneaire ou tourisme rural integré?* (Schlechten 1988), she engaged in eight uninterrupted months of fieldwork, several trips to Senegal as a tour guide, and searches of the archives and relevant literature. Since her thesis work and its publication, Schlechten has continued her interest in Africa by working for a Swiss Aid organization.

To assess the consequences for the hosts of the two different forms of tourism development in Senegal, Schlechten chose to compare mass tourism development, represented by three resorts on or near the beach of the *Petite Côte* (80–90 km south of Dakar), with the alternative *tourisme rural intégré*, represented by three villages in the bush of the less developed Lower Casamance (in the extreme south of Senegal near the border with Guinea-Bissau). Both of these tourist destinations attract a mostly European clientele, but they differ in that the villages in the Lower Casamance, as might be expected, tend to draw on younger, more adventurous types.

Schlechten used standard ethnographic procedures in the field, which included participant observation during her field work and as a tour guide, and interviews with officials, tourists and their immediate hosts. In the best tradition of ethnographic fieldwork, she used non-standardized, as well as standardized techniques. She appears to have been effectively on the ground, so to speak, with the added advantage of being able to witness tourism development through time as it was actually occurring.

The theoretical orientation of Schlechten's study derives from what might be loosely glossed as dependency theory, which often has been mentioned in this book, especially in *chapter 2*. The author sees tourism development in Senegal, whatever its specific nature, as an international phenomenon in which the destination area is dependent on, and subordinate to outside powers in Europe. All of Schlechten's criteria for analyzing and evaluating the consequences of the two forms of tourism development emerge out of this theoretical framework.

On the economic side, she raises questions about whether or not a given project tends to liberate hosts from the standard Western (mass) model of tourism development, reduce, or at least not contribute to the gap between the rich and the poor, etc. On the sociocultural side, she asks whether it tends to promote egalitarian contacts between hosts and tourists, contributes to the maintenance of a local cultural tradition, etc.

In terms of her criteria, on the face of it, tourism development on the *Petite Côte*, in which the government of Senegal has a heavy investment, would seem to have little to recommend it. The tourism there is of the large-scale, enclave type. It appears to be imposed from outside, disrupts – even overwhelms – the local culture and generates various social inequalities, superficial, commoditized relationships and an increased dependence on Europe and the outside (the big-paying tourism jobs have gone to outsiders, for example). On the other hand, the alternative in the Lower Casamance is small-scale and depends in all of its facets on the considerable involvement of the local people (it is fully staffed by villagers, all of whom have a say in its operation). Benefits are supposed to accrue to the whole village, and tourist guests, having received some orientation at home, have ample opportunity to participate in the daily life of their hosts.

On the basis of her analysis, Schlechten thinks that the alternative form of tourism *autocentré* in the Lower Casamance has much to recommend it, but she is cautious in her conclusions. Such caution is justified. In the first place, the resorts on the *Petite Côte* have hundreds of beds, while no village in the Lower Casamance has more than 40 beds. Moreover, the alternative tourists are not staying as long as was planned – most only for a day or two. The income from this alternative form of tourism, therefore, and its impact, cannot be very great. Secondly, Schlechten, herself, worries about side effects that have not been 'caught' by her comparatively short-term investigation. Thirdly, though she is at pains to sort out different kinds of developmental consequences, as, for example, those due to modernization, urbanization, or tourism development, she is not able to specify the exact part that each plays in the sociocultural changes observed (for example, though tourism development appears to contribute to the decline of the extended family through processes which she specifies, that form of family already appears to have been weakened by modernization and drought). Eventually, we should know the part that each factor plays in this change. Finally, Schlechten tends to dismiss the mass tourism development on the *Petite Côte* in a rather draconian manner that prevents her from considering that local people derive any benefits from it. For example, some of the negative effects she notes there, such as employment of outsiders in the big-paying jobs, may disappear as the local population adapts to this new enterprise. Of course, one cannot blame the author for not anticipating the current secessionist struggles which have closed down tourism there.

Though we should, perhaps, take the author's unremittingly negative view of mass tourism development in a fairly typical Third World country with a grain of salt, one is bound to admire the comprehensive nature of Schlechten's inquiry into the consequences for a less developed society of the development of forms of mass and alternative tourisms. She has produced a major work that addresses both practical and theoretical questions. It is knowledgeable about tourism and significant developments in the contemporary world, sensitive to subjects, and methodologically sophisticated. The author shows an admirable awareness of herself as investigator and a passion for the issue of social development in the Third World. To sum up, her work, which is of an applied nature, constitutes a model for all researchers on tourism development. It is unfortunate that her book has not yet been translated into English and other languages so as to be available to a broader audience.

* * *

It has been the central thesis of this chapter that basic and applied research on tourism can go hand in hand and that each is capable of enriching the other. Even as the anthropological enterprise profits from anthropologists' relations with their subjects and colleagues in and out of anthropology, it also can profit from their participation in applied research having practical ends. Such enrichment does not occur auto-matically, however. It requires the full awareness of the anthropologist involved and appropriate actions that will maintain and possibly advance the discipline. It should be emphasized that there need not be anything Machiavellian in all this. Anthropologists these days have become much more open about themselves and their research ways. They have learned to accept the participation of subject peoples in setting those goals, carrying out their procedures and even formulating conclusions. Collea-gues, as well as subjects, now know more about what researchers are up to. Why shouldn't it be the same with potential clients?

There remains one problem, however, that may not be resolvable. Applied tourism research addresses practical questions, but these ques-tions concern tourism in the contemporary world, which, according to the broadest anthropological viewpoint, is only a fraction (how big?) of all the tourism that has ever been practiced. Statements about tourism can be restricted to certain sociocultural contexts, of course, such as that in which the contemporary tourism industry operates, but in anthro-pological terms, they should also help us to form impressions about the general human condition in which something like tourism always appears to have played a part. Applied anthropology, which addresses practical questions in the contemporary world only (under our noses, so to speak), cannot contribute – or can contribute very little – to research that touches on these ultimate questions. On the other hand, some basic understanding of how humans have always operated – what is universal versus what is culture-specific – can certainly contribute to applied projects in the contemporary world.

Chapter 9

Conclusion

Retrospective

Anthropologists often have problems in explaining to others what it is, exactly, that they do, and those who study tourism are no exception. After dealing with an interlocutor's various misconceptions of anthropology and justifying the study of tourism as a part of the anthropological agenda, the anthropologist still may see signs of incomprehension and wonder whether it is worthwhile going on. Hopefully, readers who are coming to the end of this book, which has looked at tourism from an anthropological point of view, will not require much elaboration on these matters; but perhaps it would help at this point to go over the essentials of the anthropological approach and its applications to tourism. It also may be useful to summarize the important contributions of anthropology to the understanding of tourism, point out once again the pros and cons of different anthropological approaches to the subject, and speculate a little about the future of this obviously fertile field of study.

In *chapter 1* an attempt was made to get at the basics of the anthropological approach to humans and their various activities, among which are leisure, travel, and their composite, tourism. To speak of anthropology in a global sense these days, one must proceed with caution because there are so many contending points of view among anthropologists about what it is they do. For example, there are perennial contentions between the more scientific and more humanistic (interpretive) wings, between generalizers (looking for what humans in different societies have in common) and culturally specific particularists, politicizers and the apolitical, the more genetically and more culturally oriented, etc. Some have argued that the contentions are so many and so profound that anthropology may be falling apart as a discipline.

This view is too extreme. It still seems possible to throw a net that is loose and not too narrowly drawn over most, if not all of the different parties. Anthropology, or at least mainstream anthropology, can be taken as the study of humans wherever they have existed. This study,

though it has an interpretive–humanistic side, must, in the view of the
author of this book, ultimately side with science, which means attempting to develop an empirically-based, generalizable picture of the world
as it is. It is the sociocultural side of human life, that is, the life-ways of
human collectivities, which is particularly relevant for those studying
tourism.

An important feature of the anthropological study of sociocultural life
has been the problem of understanding other human beings, whose
actions, unlike those of, say, molecules, are informed by intentions and
meanings which derive from ways of life (cultures) that can be very
different from the anthropologist's own. This has mostly been accomplished by ethnographic study in the field, during which the anthropologist participates in, and questions people about their activities. It has
been traditional for anthropologists to see human actions as taking place
within sociocultural and environmental contexts and often to make
cross-cultural comparisons that may ultimately include all of humankind.
Any anthropologist, therefore, should be at home in a multicultural
environment, which, because of the travel involved, is an essential
feature of tourism. One could take a more fine-grained view of anthropology, but only by noting an increasingly unacceptable number of
exceptions.

Any research on tourism that exhibits one or more of the minimal
anthropological concerns just discussed, whether carried out by anthropologists or not, has been considered grist for the mill of the anthropological study of tourism offered in this book. So, the work of
geographers, psychologists, political scientists, economists, sociologists,
recreation and leisure scholars and others, as well as anthropologists, has
been touched on here. One can justify this by remembering that study of
tourism in anthropology is still in its formative stage where exploration is
important and boundaries are not well established. Anthropologists
studying tourism often are fairly sophisticated about work on the subject
by scholars from other disciplines. The author's own experience in the
International Academy for the Study of Tourism, which is an organization
made up of leaders in the field of tourism research from a variety of
cultures and disciplines, is a testimony to this.

The anthropological view of tourism presented in this book has been
informed by the essentials of the anthropological approach just described. The focus of the picture is the tourist, who has been conceived here
as a person at leisure who also travels. Tourists and those who have
anything to do with them (hosts, transport personnel, guides and others)
are the actors who give agency to the actions studied by anthropologists
and other social scientists. All of their activities have been seen here to
make up a multicultural touristic process which begins with the generation of tourists, continues with travels by the tourists to destination areas

where they become temporary sojourners and ends with their return home. All of this involves various kinds of social transactions, which have consequences for the people involved.

The touristic process which applies in any particular case may 'harden' into a touristic system, as is becoming evident in the developing international tourism industry today. It fits into a more or less extensive historically situated social and environmental context that shapes its nature and development. Finally, it should be remembered that it has a dynamic, change-oriented dimension, as is especially obvious in any case of tourism development. In their studies of tourism, anthropologists have paid particular attention to this diachronic dimension.

Considering that tourism now is one of the world's largest 'industries', interest in it (as well as in leisure) by social scientists seems to have been slow to emerge. Anthropologists, for reasons that were suggested in *chapter 1*, appear to have been among the slowest to catch on to the significance of the subject. But though tourism study in anthropology has been slow to develop, and though it may still be somewhat marginal in the discipline, it seems to have finally gained a measure of acceptance and even some special favor among pockets of anthropologists ever on the outlook for some new wave of interest.

In its early days, and continuing still, the anthropological study of tourism has focussed on the destination, host or supply side of the touristic process, particularly in the less developed world where many anthropologists have done their work. Early studies often were purely descriptive (with theoretically informed work being quite rare) and tended to be concerned with whether tourism was good or bad (mostly bad as far as anthropologists were concerned) for some host people. Since then, as has been the case in the other social sciences, generally, there has been a trend towards treating tourism in a more scientific way (in Jafari's terms, a change from 'cautionary or advocacy' to the 'knowledge based platform').

It has been suggested throughout this book that though there has been scientific progress in their studies of tourism, the social sciences still have a good way to go in studying it with the combination of methodological competence and theoretical sophistication that is the mark of a mature science. Anthropology may be even farther removed from such a state than the other social sciences because of ideological leanings towards muckraking, as in many studies of the consequences of contact of less-developed peoples of the world with the more developed, and a tradition of subjective interpretivist–humanistic analysis, as, for example, in the works of famous anthropologists like Ruth Benedict (see e.g., 1946) and Clifford Geertz (see e.g., 1960), who have attempted to get at the essentials of particular cultures through their interpretive art.

Still, it would be a mistake not to recognize the contributions of anthropological studies to our understanding of tourism. Looking only at the destination end of the touristic process, which is where most anthropological interest in tourism has been concentrated, one can see a good deal of progress in one aspect of the anthropological agenda, which is to chronicle the ways of different cultures of humankind. Thanks to anthropologically-oriented researchers, we know a good deal more about different forms of tourism development, particularly in the less developed world. A glance over the various case studies of tourism development in Smith's edited *Hosts and Guests* (1989), which may be the best-known book in the field, will attest to this. Moreover, some progress is evident in sorting out, or classifying these various forms of tourism development.

In contrast to this descriptive aspect of the scientific task, there is less progress to report on the analytic–explanatory side. Science is concerned with processes, but in tourism studies by anthropologists, processes (such as demonstration and ripple effects) are referred to, if at all, only in a rudimentary way. Science is also concerned with explanation, which involves the search for various kinds of causes and effects. As far as tourism development is concerned, some progress has been made in separating out tourism from other possible causal agents (e.g., industrialization, urbanization, non-touristic migration) of developmental changes in host societies, but methodological inadequacies often mar such studies. Too often, it is simply *assumed* that tourism is the cause of some concomitant change in the host culture.

Explanations in science involve recourse to grander or lesser theories, which are general statements of how different parts of the world are related to each other. Looking at the subject of tourism development only, which has preoccupied anthropologists, it is not easy to find theoretically sophisticated work by anthropologists, whether the theory is derived from anthropology or other disciplines. This state of affairs, which also exists in the field of leisure studies, has been lamented, perhaps most notably by Greenwood. One wonders whether this under-theorized condition of tourism study in anthropology has anything to do with its marginality in the discipline.

Connectedness with some regnant body of theory certainly is a way of demonstrating belonging. Whether a theory 'works' or not may not be as important for its respectability in some discipline as whether a study simply has some theoretical relevance. It is obvious that the anthropological study of tourism needs a better theoretical underpinning, which, of course, has to be combined with many more solid ethnographic investigations on the ground in a variety of cultures.

The progress which anthropologically oriented researchers have made in describing the various forms of tourism development in different cultural settings is not matched by that in comprehending other facets of tourism, but other perspectives on tourism do exist. One can discern three general perspectives that anthropologists have used for viewing the subject. The first, which has dominated the field and which has guided studies of tourism development, has seen tourism as a form of acculturation or development. Here, the emphasis has been on what has happened to host peoples, particularly in the less developed world, as the result of the introduction of tourism. This approach and the work associated with it was discussed in *chapter 2*.

The second general view of tourism taken by anthropologists, which was given a particularly critical examination in *chapter 3*, has focused on tourists and their odysseys. It looks at tourism as a personal transition. Tourists in their passages are seen to be moving from one experiential state to another, with greater or lesser consequences for them (and possibly for their home societies).

The third general anthropological perspective on tourism, which was discussed in *chapter 4*, is barely evident in the literature, but, because it deals with the causes of tourism, may have the potential to be the most significant of all. Taking a perspective that amounts to viewing tourism as a kind of social superstructure, a few researchers have looked at the processes by which tourists and tourisms are generated, in which the home society figures importantly. This line of investigation, as well as that which has looked into the consequences of tourism development for host societies, raises, among other questions, the issue of how different aspects of a sociocultural system are related to one another.

These three general perspectives (not exhausting the possible ways of looking at tourism), especially when taken in the context of the broadened concept of acculturation offered in *chapter 5*, would seem to go a long way in comprehending the touristic process, which is to say, tourism in all of its aspects; and after appropriate revision, they would appear to provide a good foundation for anthropologists and other social scientists to carry on future work in the field.

Criticism of anthropologically oriented research that has been carried out from the three general perspectives was included in *chapters 2, 3, 4* and subsequently, *chapter 5*. Going beyond the comments about the first perspective, that looking at tourism as acculturation or development, which have been offered earlier, consider the view of tourism as a personal transition, which has been dominated by a model derived from Turner's and, ultimately, van Gennep's work on ritual processes, including pilgrimage, as adapted for the study of tourism by Graburn. In passing through a ritual-like process, the initiate (tourist) has been seen by some anthropologists to be moving from a structured to a destructured experiential state and back again. Aside from the difficulties in identifying these states empirically, as well as the alleged need for alternation or inversion of experience in human beings that is said to generate such a transition, there is the question of how many tourists conform, more or less, to the model. Only a little experience with the subjective side of tourists' odysseys leads one to believe that only some tourists (probably a very few) do so. It does look as though the experiences of tourists vary widely, and even if a need for inversion were demonstrated to exist, it would be swamped as an explanatory factor by any one or more of a variety of other factors (all of them related to some social context or other) that were considered in *chapter 3.*

One of the more important of these other factors, pointed to by Turner's notion of *communitas,* is the tourist peer group, which would be a congenial subject for many future anthropological investigations of tour group cultures. Questions of how their touristic experiences affect individual tourists and their home societies, implicit in the research agenda of tourism as a personal transition, have only begun to be explored.

From the third perspective, which views tourism as a kind of superstructure, that is, as one of the more dependent aspects of a sociocultural system, one can look into the tourist generating situation, which includes the home society of tourists. Anthropologists may not feel comfortable investigating a tourist generating situation, which may involve their own more developed society, but it would seem that, as far as the study of tourism is concerned, this (because it treats the causes of tourism) may be the most important project of all. In this project, it often will be necessary to deal with larger scale societies, with which colleagues in sociology and other social sciences probably may be more familiar from a scientific point of view, but because tourism seems to be universal, the study of the generation of tourism in less developed societies should also be on the agenda. In *chapter 5,* after reviewing the important features of the anthropological approach to tourism, a broad view of tourism that accommodates the three general perspectives that have guided the anthropological study of tourism to date was offered. That view involves a broadening of the acculturation paradigm that has guided a good deal of anthropological work on social change in the past. It is, in fact, an elaboration on the concept of touristic process which has provided the basic framework for viewing tourism in this book. According to this broader view, tourism may be seen to involve contact between agents of historically situated tourism generating and receiving situations with consequences for all parties involved in that contact. Any subject or research problem about tourism can to be located somewhere in this larger picture and approached from the slant of one of the three perspectives that have governed anthropological research on tourism to date, or possibly some other, using theory that fits the issues involved.

In looking at the applied side of anthropologically-oriented research on tourism (*chapters 6, 7* and *8*), it was argued that it is essential to think first about some tourism field in the touristic process, in which a variety of actors, each of them pursuing their own interest, are involved. Any touristic policy and its implementation may be seen as emerging out of the give and take between the interests of the actors who have some stake in a tourism project or event. So, a host government, its people, an outside consulting agency (with, perhaps, an anthropologist on board) and some multinational hotel chain may be involved in setting the policy for, and implementing the development of a particular resort. Any applied anthropologist involved in such a tourism project should, routinely, be aware of its social field so as to consider the possibilities for realization of different alternatives.

Chapter 6 also discussed what applied tourism researchers can do (helping with the development of policy, implementing that policy, monitoring, and assisting with mitigation) and what they actually do. Generally, it appears that applied researchers have realized only a small fraction of their potential for assisting in tourism projects, and this certainly is the case for anthropology. Most of what has been accomplished fits in the category of impact analysis, which makes an assessment (either before or after the fact) of the impact of tourism development on a host society and its environment. In this regard, the concept of sustainable development has come to the fore. In *chapter 7*, it was pointed out that though there appears to be a widespread acceptance of this concept, it needs to be more fully understood and critically evaluated. To that end, an extended discussion of ecotourism as a sustainable alternative and the Bali Sustainable Development Project, which has an important touristic component, were included in this chapter that concluded with suggestions about how the concept of sustainable development could be used in applied anthropological work on tourism.

Applied research can contribute to the anthropological enterprise if for no other reason than its provision of jobs for anthropologists. It can also introduce anthropologists to what may turn out to be fertile fields of investigation, as, for example what goes on in board rooms of organizations developing tourist resorts. There are problems, however, in linking up with parties that have vested interests in particular courses of touristic action. Ideally, the anthropologist should be governed by a purely scientific agenda, but this, of course, is hardly ever the case. A variety of more or less recognizable social and psychological factors can shape scientific constructions of reality. In applied research, one of those factors clearly involves a client's interests, which may be at odds with an anthropologist's more basic scientific agenda.

In *chapter 8*, the question of whether basic and applied anthropological interests can coexist was raised and (granted that certain cautions are observed) answered in the affirmative. Various courses of action usually are available to researchers whose personal and disciplinary interests conflict with those of their clients. An awareness of these possibilities can be heightened by some anthropological consciousness raising, which would include understanding the nature of the applied project, the relevant field of social action involved, including the roles of researcher and client in it, and of oneself. A discussion of some problems of cultural consultants in heritage tourism projects was included in the chapter to illustrate the relevant issues; and to conclude, a number of cases in which applied anthropological work could inform, or be informed by more basic scientific concerns were discussed.

One has to keep in mind, though, that applied research has not, or cannot be applied to all touristic subjects. Applied anthropologists will never be able to attend to the problems of touristic actors in historic, nor pre-historic societies. This means that, though there can be constructive give and take between applied and more basic anthropological interests, the applied contribution to an ultimate anthropological agenda that concerns all of humankind will be somewhat limited. However, work on that ultimate agenda, because it deals with what is universal and what is culturally specific, can contribute to the work of the applied anthropologist.

What Tourism Now Looks Like

At the outset of their efforts to comprehend tourism, some anthropologists may have thought that they were on to an easily identifiable and manipulable subject. Only a little acquaintance with the manifold variety of tourism should have disabused them of such a notion. First, there was (and is) the problem of what tourism is. Then, after settling on a definition, it quickly became evident that tourism takes many different forms, especially when it is viewed from the cross-cultural perspective that informs anthropology.

All of this should not be surprising for anyone acquainted with the history of anthropological thought. Take, for example anthropological inquiry into the subject of religion. There have been different notions about what religion is, whether it can be distinguished in a given instance from other areas of social action, and whether the Western-based term can be used for inquiring about the subject in cultures which do not set it aside conceptually. Nevertheless, despite these problems, two more or less explicit notions of religion have dominated the field. These see religion centering around some concept in the minds of people that can be construed as either sacred or supernatural. As far as anthropologists are concerned, therefore, there is no single notion of religion, but at least two.

It is obvious, also, that religion has a number of facets, which have been seen to include belief, ritual and social organization, among other things.

In their studies of religion, anthropologists have sometimes tried to include all of these facets, but often have concentrated on one or the other. It is obvious, then, that as far as the entire discipline is concerned, religion has different meanings or facets. Added to this, of course, is the fact of intracultural and transcultural variation (polytheism, monotheism, animism, for example) which may undermine preferred views of the subject.

As far as the field of tourism is concerned, the situation is not very different. From anthropological explorations of the subject, there have emerged two major definitional traditions, both of which are based on some notion of the tourist, who has been conceived primarily either as a person at leisure or a traveler. Tourism, then, refers to the actions (or transactions) of these tourists and those who serve or deal with them. Such definitions, as with all scientific definitions, inevitably encounter problems of applicability in a multifarious world, but, if some fuzziness around the edges is tolerable, these problems can be resolved. However, researchers employing the same definition will see things differently depending on which actor in the touristic system is the focus of their investigation.

Furthermore, though their views can be fitted together into a broadened acculturation perspective, anthropologists looking at tourism from the perspective of acculturation or development obviously will see it differently from those who see it to be a kind of superstructure or as a personal transition. So far, the main picture of tourism that anthropology offers has emerged from the application of the acculturation or developmental perspective in which hosts in some destination area, particularly in the less developed world, play the leading role. Host peoples are seen to be benefiting or (often) not from tourism, being subjected to Westernizing influences and a drift towards dependence, experiencing more or less economic leakage, having favorable or unfavorable attitudes towards tourists and tourism, etc. The picture that has emerged is a dynamic one in which developing tourism practices are seen to be associated with a variety of sociocultural changes among the hosts through processes such as the demonstration or ripple effect.

The limitations of this view of tourism and its inadequate cross-cultural coverage should be evident. There are other facets of tourism and other contexts involving other cultures that need to be considered in drawing up a full picture of tourism from an anthropological point of view. Further, though anthropologists have made progress in looking at tourism in a more scientific way, they still have a way to go towards scientific maturity. This means that one needs to be cautious about the empirical basis of the pictures of tourism they have offered. Still, one can be grateful for the light they have thrown on the subject, particularly in areas where other scientists have less interest and expertise.

As far as analysis and lines of explanation are concerned, anthropologists studying tourism have not made as much progress as those studying, say, religion, where theories have proliferated. In the study of religion, there have been plenty of more or less (usually less) well-grounded theories to go round. In the anthropological study of tourism, on the other hand, research has, with a few important exceptions, rarely been theoretically informed. Moreover, in those rare studies where theory has been deployed in a consequential way, there usually have been significant methodological deficiencies. The struggle to come to terms with the 'other' always looms important in anthropological work, but it cannot be used to excuse methodological and theoretical inadequacies, which, sooner or later, must be addressed if the anthropological study of tourism is to be taken seriously as a scientific endeavor.

Into the Future

Despite its deficiencies, so much has been accomplished and so much is in place in the field of tourism studies in anthropology that one can be optimistic about the future. In the first place, such research has the potential to raise all of the significant questions that have occupied anthropologists and other social scientists; and because it is a comparatively new field of study, these questions can be looked at in a somewhat different light, which usually is valuable for scientific progress.

Moreover, because the anthropological perspective is so broad and all-encompassing and because there is so much cross-fertilization with other disciplines, it can provide a unifying vision for the various social scientific approaches to the subject. That vision involves the concepts of touristic process and touristic system, as well as the associated expanded acculturation perspective that was developed in *chapter 5*. Moreover, the three interrelated perspectives by which anthropologists have studied the subject give specific directions for anthropological and other inquiries. In sum, it seems fair to say that the emerging field, though certainly not without its deficiencies and still involving a good deal of clearing away of unproductive clutter, is soundly based and on the right track. It is to be hoped, however, that certain steps will be taken so that the anthropological study of tourism will fulfill its promise.

First, it is to be hoped that increasing numbers of anthropologists and colleagues in other disciplines will find tourism to be a legitimate subject to be investigated. Though more and more scholars are coming on board, one still wonders from time to time whether people in the mainstream of anthropology have been fully convinced of the subject's legitimacy and if they hold views similar to those of an international sample of leisure and recreation scholars surveyed by Jackson and

Burton (1989:14–15). These scholars tended to feel that anthropology has contributed very little to their fields of study. Would anthropologists respond similarly in regard to tourism studies by their colleagues? In anthropology, there is a legacy of accepted areas of investigation – development, religion, kinship, economic activity, political organization and language, among them. For a young Ph.D. to offer tourism as an area of specialization these days might add little to the value of an application for an academic position in some Western or Western-like university. It is possible, of course, to provide a number of justifications for the study of tourism, among which are its importance as an industry, its link with issues of development, etc., but despite the many changes in the field of anthropology, there are certain traditions that die hard. It may still be necessary for some time longer for young anthropologists concerned with tourism to continue to 'piggy back' that interest on other, more established interests (development, for example).

Second, as Crick has pointed out, many more ethnographic studies of tourism, which provide the necessary raw materials of sociocultural anthropology, need to be done. Such studies, of course, should be more competent, methodologically and sophisticated, theoretically. It goes without saying that in these investigations anthropologists should continue to fulfill the anthropological requirement of giving the point of view of subjects their due, whether these subjects are in the board rooms of tourism development firms, in shanty towns near some tourist destination, or cruising on the Sepik River in New Guinea. This means, among other things, the continuation of the practice of field work in the best anthropological tradition. It also means putting researchers and their procedures into the picture so that all, including researchers themselves, will know something about how a study is oriented. This kind of reflexiveness should never be an exercise in itself, as has happened in some ethnographic works, but should always serve as a means of clarifying the picture of whatever aspect of tourism is being studied.

Third, there is a need for a broader, better balanced coverage of the touristic process wherever it exists. This means that more studies involving destination areas outside of the traditional anthropological stamping ground in the less developed world should be undertaken. So, there should be studies of tourism development in places like Glasgow, Scotland, Willimantic, Connecticut (U.S.A.), or Stockholm, Sweden, as well as in The Gambia, Nepal, or Barbados. At the same time, there should be more attention directed at other areas of the touristic process beyond destination areas. The generation of tourists and the tourists themselves should become more important foci of anthropological interest. Finally, the historical study of tourism, which poses problems of investigation related to the study of people at a distance, different kinds of source material, etc., should become routine in the anthropology of tourism. All such research, as well as that which has been accomplished so far, can be accommodated within the conceptual framework suggested in this book.

Fourth, the theoretical side of tourism study by anthropologists should be upgraded. Though it seems unlikely now that there can be a general theory of tourism, there exist a number of middle-range theories applicable to different aspects of tourism in different contexts. For example, an upgraded dependency theory, which sometimes has been assumed to apply in all questions concerning the effects of tourism on host societies, should certainly be a theory of choice for destination areas in the less developed world such as the Caribbean, but it would seem to be less relevant in places like Paris. Theories of commoditization, bureaucratization, or 'technologization', can be wedded to some variant of world-system theory in studies of the causes and consequences of tourism generated in the huge, impersonal societies of today's world and all they touch, but they are not universally applicable. As we have seen, the use of theory derived from analyses of the ritual process has to be operationalized and applied selectively to tourist experiences and reactions in particular contexts. There are also a variety of theories of leisure, which, if appropriately formulated for tourism specifically, can inform questions concerning the generation of tourism or its function in a society. All of these theories can be applied within the broad acculturation perspective suggested here.

The theoretical state of anthropology and neighboring disciplines in the social sciences is largely beyond the purview of this book. However, we do know that there are a variety of theories available that can be tailored to anthropological work on tourism. For example, Rojek (1985) has found a rich theoretical lode in sociology that pertains to leisure, which seems well adapted to the treatment of tourism as a kind of superstructure. It would seem that the time has come for an understanding among anthropologists and other social scientists that every study of tourism should have some explicit or implicit theoretical point and that it should play some role in theoretical advances in the discipline. This means that anthropologists, besides having a transcultural understanding of the touristic process, should have some idea of the range of tourism-relevant theories available in anthropology and other disciplines. It also means that competing theories should be tried on with care and tested with appropriate methodological procedures.

Though there is not likely to be a single theory of tourism in the foreseeable future, there can be a conceptual scheme that will embrace it all. As Nash (1992:225) points out, such a scheme is ready to go. This scheme, which has been developed to meet anthropological concerns, is provided by the concepts of touristic process or system and the broadly conceived notion of acculturation which have dominated this book. Though this scheme is especially adapted to the anthropological considerations, it also can be used as an integrating device for the study of tourism by all of the social sciences. It provides a general overview of touristic phenomena that lends coherence to investigations which each discipline can pursue from its own point of view.

Fifth, anthropologists investigating tourism have, so far, shown a laudable open-mindedness in their work. Though some institutionalization is bound to set in with the development of their field, they should seek to remain open to various social influences and profit from the dialogues involved. As has been suggested earlier, they can benefit not only from dialogues with their subjects, but also among themselves (between applied and basic scientists, for example), with other social scientists studying tourism and with a growing number of colleagues from other cultures in a variety of disciplines. Hopefully, all of this give and take will lead to more viable points of view and procedures for investigating tourism in different contexts.

It goes without saying, of course, that for a discipline with comparative, cross-cultural concerns such as anthropology, the recruitment of more and more researchers from other cultures outside of the Western core area where the anthropological study of tourism got its start and in which most of its practitioners continue to operate is absolutely essential. These new colleagues will not only add new dimensions to the picture which anthropologists can present, but they also will help in gaining the empathy with 'others' that will continue to be an essential part of the anthropological game.

Despite accusations leveled against anthropology as a Western 'hegemonic' science, the author of this book feels that, if special problems of anthropological investigation are dealt with, it remains a superior way of knowing the world of human beings. Accordingly, contributions of these additional non-Western scholars to the understanding of 'others' will have to be matched by some commitment on their part to the basic tenets of anthropology and to empirical science which have been discussed earlier in this book. The aim must be not only an accumulation of points of view about tourism. Eclecticism should never be an end in itself, but rather, a means for developing better and better collective perspectives for grasping tourism realities. Needless to say, this will involve lively debate, which is a hallmark of any vigorous science.

* * *

Hopefully, the study of tourism by anthropologists and like-minded colleagues will continue to gain momentum along lines that have been sketched out here. Anthropologically-oriented scholars will then be carrying out their scientific mandate to explore cross-culturally all aspects of tourism, to search out its causes and look for its consequences. Because the subjects of tourism study are human beings who come from many different cultures, the problems of studying them scientifically, which have been pointed to throughout this book, should constantly be attended to. These problems include understanding a wide variety of human beings, who often are products of cultural milieux other than that of the researcher, and attending to anthropologist-observers and their procedures as factors affecting the picture of tourism being developed.

The general outlines of the multifarious picture of tourism which anthropologists are developing are gradually emerging. There is, however, still another element in their view that should be mentioned. This has to do with the concept of culture, which informs anthropological views of human action. The notion of culture is so much taken for granted by anthropologists and like-minded brethren in other disciplines, as well as by increasing numbers of the general public, that it has been touched on only lightly in this book. But now, as this anthropological study of tourism comes to an end, it might be helpful to consider the concept more fully and point out how it contributes to the anthropological picture of humankind and, more specifically, of their touristic activities.

The term 'culture' has been used in a number of ways. For anthropologists, it refers, first of all, to human actions. Unlike one other usage, which reserves the term for a certain kind of activity such as art or intellectual discourse, the anthropological reference is to the whole panorama of human behavior of which such activities are only a part. Different aspects of human action may be stressed, as, for example the ideational as opposed to the behavioral, but always human actions are seen to be of equal value, and so, are to be studied with the same seriousness of purpose. From this point of view, tourism becomes one of a variety of human activities that invite serious scientific scrutiny.

Secondly, anthropologists use the term culture to refer to the learned component of action, which, in humans, as compared to other animals, is very great. Because of their great learning ability, humans are comparatively free of biological dictates in developing courses of action. They are not totally free of biology, of course, but their actions are dramatically different in the learned dimension from those of, say bees, whose behavior is mostly programmed by genetic mechanisms that change quite slowly in a manner independent of changes in the environment. The concept of culture celebrates the flexibility of human action which anthropologists have chronicled in their cross-cultural studies and their investigations of sociocultural change.

By thinking in terms of this meaning of the culture concept, the potential of humans to change their ways would seem to be very great indeed. Thus, it would not seem inevitable that women in some society (as compared with men) should suffer so many deprivations; and there would be hope that the underclasses of the world could overcome oppression and their self-defeating ways and rise up to control their destinies. Where tourism is concerned, alternatives to some current unpalatable practices would become real possibilities. From this perspective, the herd-like tendencies of mass tourism, the exploitive character of tourism in the Third World, the superficial relationships between tourists and hosts – all could be seen as capable of being made more acceptable to the people involved.

There are very few, if any anthropologists who share this concept of culture and all that it implies, who would be so naively optimistic about human flexibility. This is because, for them, culture has still another meaning, which refers to the normative tendencies that exists in all group actions. According to this usage, the weight of social custom is stressed and culture becomes a socially constraining or directing force on the people of some group. One speaks now of the customary practices of groups such as hunting and gathering, agriculture, Christianity, etc. People, because they must live in groups to survive, have to acquire (through learning) ways of acting that will permit them to get along with other group members. The fit is never perfect and human individuality always is maintained, but there are behavioral limits in all groups, beyond which people run the risk of becoming *persona non grata*. So, Theravada Buddhists are Theravada Buddhists in their actions and not Mormons because Theravada Buddhism is the religion of their group. Using this version of the culture concept, one can see, where tourism is concerned, that terrorists attack tourists in a certain country, that certain developers pursue their work without thinking about social and environmental costs, or that aboriginal San in a certain band tended to favor another band for their touring – all 'because of their culture'. Seen in this way, tourism or any other kind of human activity becomes action that is dictated by socially acquired group rules or norms, which, though capable of change, would seem to undercut the great potential for change implied in the second meaning of culture.

There is still another way in which the concept of culture, as used by anthropologists, may be considered. According to this usage, the possibilities for changes in human activities are to be seen in terms of the workings of culture, which it is the task of the anthropologist to comprehend and communicate to others. From this point of view, tourism can be viewed as conforming to laws about which we now know all too little, but will certainly know more. Anthropologists, as their scientific work on tourism advances, will be able to contribute more and more to our understanding of these laws. By becoming better acquainted with different aspects of tourism in a broader and broader range of human societies, increasingly knowledgeable about its workings, more aware of views of a multitude of subjects and colleagues from a variety of disciplines and cultures, more sophisticated, theoretically and competent, methodologically, continuing their preferred position of involved outsider and increasingly aware of themselves and their role in the research process, they should be in a privileged position for illuminating what people gradually are coming to realize is a significant aspect of the human condition.

References

Acero, J., and C. Aguirres
1994 A Monitoring Research Plan for Tourism in Antarctica. Annals of Tourism Research 21(2):295–302.

Adams, A.
1993 Dyke to Dyke: Ritual Reproduction at a U.S. Men's Military College. Anthropology Today 9:3–6.

Adams, R.
1974 Anthropological Perspectives on Ancient Trade. Current Anthropology 15:239–258.

Adams, V.
1992 Tourism and Sherpas, Nepal: Reconstruction of Reciprocity. Annals of Tourism Research 19:534–554.

Ahmad, J.
1985 Lost in the Crowd. Washington DC: Three Continents Press.

Ahmed, A., and C. Shore, eds.
1995 The Future of Anthropology: Its Relevance to the Contemporary World. London: The Athlone Press.

Albers, P.
1992 Postcards, Travel and Ethnicity: A Comparative Look at Mexico and the Southwestern United States. Paper presented at the Meetings of the American Anthropological Association, San Francisco, CA.

Albers, P., and W. James
1983 Tourism and the Changing Photographic Image of the Great Lakes Indians. Annals of Tourism Research 10:123–148.
1988 Travel Photography: A Methodological Approach. Annals of Tourism Research 15:134–158.

Anastasopoulos, P.
1992 Tourism and Attitude Change: Greek Tourists Visiting Turkey. Annals of Tourism Research 19:629–642.

Ap, J.
1990 Residents' Perceptions Research on the Social Impacts of Tourism. Annals of Tourism Research 17:610–616.

Aspelin, P.
1978 Indirect Tourism and Political Economy: The Case of the Mamaindê of Mato Grosso, Brazil. In Tourism and Economic Change. Studies in Third World Societies No. 6. Williamsburg, VA: College of William and Mary, Department of Anthropology.

Balsdon, J.
1969 Life and Leisure in Ancient Rome. New York: McGraw-Hill.

Barth, F.
1961 Nomads of South Persia. Oslo: Oslo University Press.

Barthes, R.
1972 Mythologies. New York: Hill and Wang.

Bee, R.
1974 Patterns and Processes. New York: The Free Press.

Benedict, R.
  The Chrysanthemum and the Sword. Boston: Houghton Mifflin.
Bidney, D.
  1967 Theoretical Anthropology (2nd ed.). New York: Schocken Books.
Bird-David, N.
  1992 Beyond the Original Affluent Society: A Culturalist Reformulation. Current Anthropology 13:25–48.
Blangy, S., and T. Nielsen
  1993 Ecotourism and Minimum Impact Policy. Annals of Tourism Research 20:357–360.
Bochner, S.
  1977 Friendship Patterns of Overseas Students: a Functional Model. International Journal of Psychology 12:277–297.
Boissevain, J.
  1978 Tourism and Development in Malta. *In* Tourism and Economic Change. Studies in Third World Societies No. 6. Williamsburg, VA: College of William and Mary, Department of Anthropology.
Boorstin, D.
  1964 The Image: A Guide to Pseudo-Events in America. New York: Harper and Row.
Borowsky, R., ed.
  1994 Assessing Cultural Anthropology. New York: McGraw-Hill.
Botterill, T.
  1991 A New Social Movement: Tourism Concern, the First Two Years. Leisure Studies 10:203–217.
Bourdieu, P.
  1984 Distinction. London: Routledge and Kegan Paul.
Boyer, M.
  1972 Le Tourism. Paris: Éditions du Seuil.
Brackenbury, M.
  1993 Trends and Challenges Beyond the Year 2000. Round Table on Tourism Trends and Challenges Beyond the Year 2000. Bali, Indonesia: 10th General Assembly, World Tourism Organization.
Brislin, R.
  1981 Cross-Cultural Encounters. New York: Pergamon Press.
Brown, B.
  1985 Personal Perception and Community Speculation: A British Resort in the 19th Century. Annals of Tourism Research 12:355–370.
Brown, D.
  1991 Human Universals. Philadelphia, PA: Temple University Press.
Brown, G.
  1992 The Meaning of Tourist Experiences. *In* Joining Hands for Quality Tourism: Interpretation, Preservation and the Travel Industry (Proceedings of the Heritage International Third Global Congress, Honolulu, Nov. 3–8, November, 1991), R. Tabata, J. Yamashiro and G. Cherem, eds., pp. 43–44. Honolulu: University of Hawaii, Sea Grant Extension Service.
Bruner, E.
  1991 Transformation of Self in Tourism. Annals of Tourism Research 18:238–250.
BSDP (Bali Sustainable Development Project)
  1990 Report on the Second Workshop: Environment and Development in Bali. Sanus, Bali: University Consortium on the Environment.

1991a Report on the village of Kelurahan Kerobokan (Research Paper 12). Waterloo, Ontario: University of Waterloo, Faculty of Environmental Studies.

1991b Report on the Village of Bongkasa (Research Report 15). Waterloo, Ontario: University of Waterloo, Faculty of Environmental Studies.

1992 Sustainable Develoment Strategy for Bali. Yogyakarta, Java: Gadjah Mada University; Waterloo, Ontario: University of Waterloo; Denpasar, Bali: Udayana University.

Butler, R. W.

1980 The Concept of a Tourist Area Cycle of Evolution: Implications for Management of Resources. Canadian Geographer 24:5–12.

1989 Tourism and Tourism Research. *In* Understanding Leisure and Recreation: Mapping the Past, Charting the Future, E. Jackson and T. Burton, eds., pp. 567–596. State College, PA: Venture Publishing.

1993 Pre- and Post-Impact Assessment of Tourism Development. *In* Tourism Research: Critiques and Challenges, D. Pearce and R. Butler, eds., pp. 135–155. New York: Routledge.

Butler, R. W., and G. Wall

1985 Themes in Research on the Evolution of Tourism. Annals of Tourism Research 12:287–296.

Byron, R.

1982 The Road to Oxiana. New York: Oxford University Press.

Caldwell, L.

1992 Globalizing Environmentalism: Threshold of a New Phase in International Relations. *In* American Environmentalism: The U.S. Environmental Movement, 1970–1990, R. Dunlap and A. Mertig, eds., pp. 63–76. New York: Taylor and Francis.

Carson, R.

1962 The Silent Spring. Boston: Houghton Mifflin.

Chick, G.

1986 Leisure, Labor, and the Complexity of Culture: An Anthropological Perspective. Journal of Leisure Research 18:154–168.

Church, A.

1982 Sojourner Adjustment. Psychological Bulletin 91:540–563.

Clawson, M.

1964 How Much Leisure, Now and in the Future? *In* Leisure in America: Blessing or Curse?, J. Charlesworth, ed., pp. 1–20. Philadelphia: American Academy of Political Science.

Cleveland, D.

1994 Can Science and Advocacy Coexist: the Ethics of Sustainable Development. Anthropology Newsletter 35(3):9–10.

Coelho, G.

1962 Personal Growth and Educational Development through Working and Studying Abroad. Journal of Social Issues 18:55–67.

Cohen, E.

1974 Who is a Tourist?: A Conceptual Clarification. Sociological Review 22: 527–55.

1979a The Impact of Tourism on the Hill Tribes of Northern Thailand. Internationales Asienforum 10:5–38.

1979b A Phenomenology of Tourist Experiences. Sociology 13:179–202.

1984 The Sociology of Tourism: Approaches, Issues and Findings. Annual Review of Sociology? 10:373–392.

1988a Traditions in the Qualitative Sociology of Tourism. Annals of Tourism Research 15:29–46.

1988b Authenticity and Commoditization in Tourism. Annals of Tourism Research 15:371–386.

1993 The Study of Touristic Images of Native People: Mitigating the Stereotype of a Stereotype. *In* Tourism Research: Critiques and Challenges, D. Pearce and R. Butler, eds., pp. 36–69. London: Routledge.

Cohen, E., ed.

1979 The Sociology of Tourism. Annals of Tourism Research 6:1–2.

1985 Tourist Guides: Pathfinders, Mediators, and Animators. Annals of Tourism Research 12(1).

Cohen, E., and R. Cooper

1986 Language and Tourism. Annals of Tourism Research 13:533–564.

Cooper, W.

1989 Some Philosophical Aspects of Leisure Theory. *In* Understanding Leisure and Recreation: Mapping the Past, Charting the Future, E. Jackson and T. Burton, eds., pp. 49–68. State College, PA: Venture Publishing.

Crick, M.

1989 Representations of International Tourism in the Social Sciences: Sun, Sex, Sights, Savings, and Servility. Annual Review of Anthropology 18:307–344.

Crompton, J.

1992 Structure of Vacation Destination Choice Sets. Annals of Tourism Research 19:420–434.

Csikszentmihalyi, M.

1975 Beyond Boredom and Anxiety. Washington: Jossey Bass.

1981 Leisure and Socialization. Social Forces 60(2):332–340.

Cukier-Snow, J., and G. Wall

1993 Tourism Employment: Perspectives from Bali. Tourism Management June:195–201.

Dann, G.

1981 Tourist Motivation: An Appraisal. Annals of Tourism Research 8:187–219.

Dann, G., D. Nash, and P. Pearce

1988 Methodology in Tourism Research. Annals of Tourism Research 15:1–28.

Dashefsky, A., J. De Amicis, B. Lazerwitz, and E. Tabory

1992 Americans Abroad: A Comparative Study of Emigrants from the United States. New York: Plenum Press.

de Kadt, E.

1992 Tourism Alternatives: Potentials and Problems in the Development of Tourism, V. Smith and W. Eadington, eds., pp. 47–75. Philadelphia: University of Pennsylvania Press.

de Kadt, E., ed.

1979 Tourism: Passport to Development? Oxford: Oxford University Press.

de Lumley, H.

1969 A Paleolithic Camp at Nice. Scientific American, May: 42–50.

Dewar, R.

1984 Environmental Productivity, Population Regulation, and Carrying Capacity. American Anthropologist 86:601–614.

Diecke, P.

1993 Tourism and Development Policy in the Gambia. Annals of Tourism Research 20:423–449.

Dollars and Sense.

1978 Tourism Travels in the Third World. Dollars and Sense 36, 14–15.

Dorcey, A.

1991 Water in Sustainable Development: From Ideal to Reality. *In* Perspectives on Sustainable Development in Water Management: Towards Agreement in the Fraser River Basin, A. H. J. Dorcey, ed., pp. 1–16. Vancouver: Westwater Research Center, University of British Columbia.

Doxey, G.

1975 A Causation Theory of Visitor–Resident Irritants: Methodology and Research Inferences. Proceedings, Sixth Annual Conference, Travel and Research Association, pp. 195–198. San Diego, CA.

Dufour, R.

1977 Les Myths du Loisir/Tourisme. Aix-en-Provence: Centre des Hautes Études Touristiques.

Dumazedier, J.

1968 Leisure. International Encyclopedia of the Social Sciences 9:248–253.

Dunlap, R.

1992 Trends in Public Opinion Toward Environmental Issues: 1965–1990. *In* American Environmentalism: The U.S. Environmental Movement, 1970–1990, pp. 89–116. New York: Taylor and Francis.

Dunlap, R., and A. Mertig

1992 The Evolution of the U.S. Environmental Movement from 1970 to 1990: An Overview. *In* American Environmentalism: The U.S. Environmental Movement, 1970–1990, R. Dunlap and A. Mertig, eds., pp. 1–10. New York: Taylor and Francis.

Durkheim, E.

(1947 [1915]) The Elementary Forms of the Religious Life. Glencoe; IL: The Free Press.

Eade, J.

1992 Pilgrimage and Tourism at Lourdes, France. Annals of Tourism Research 19:18–32.

ECTWT (Ecumenical Coalition on Third World Tourism)

1986 Third World Peoples and Tourism: Report of the 1986 Conference Proceedings. Chorakhebua, Bangkok: ECTWT.

Ehrentraut, A.

1993 Heritage Authenticity and Domestic Tourism in Japan. Annals of Tourism Research 20:262–278.

Erisman, M.

1983 Tourism and Cultural Dependency in the West Indies. Annals of Tourism Research 10:337–362.

Errington, F., and D. Gewertz

1989 Tourism and Anthropology in a Post-Modern World. Oceania 60: 37–54.

Estrada, A.

1973 Los Vuelos Charter: Su Nacimiento y Evolucion en España. *Estudios Turisticos* 40:115–128.

Evans, N.

1986 The Tourism of Indian California: A Neglected Legacy. Annals of Tourism Research 13:435–450.

Evans-Pritchard, D.

1989 How "They" see "Us": Native American Images of Tourists. Annals of Tourism Research 16:89–105.

Fanon, F.
1967 The Wretched of the Earth. Harmondsworth: Penguin.
Farver, J.-A.
1984 Tourism and Employment in the Gambia. Annals of Tourism Research 11:249–266.
Foster, G.
1986 South Seas Cruise: A Case Study of a Short-Lived Society. Annals of Tourism Research 13:215–238.
Fox, R., ed.
1995 Special Issue: Ethnographic Authority and Cultural Explanation. Current Anthropology 33(1).
Francisco, R.
1983 The Political Impact of Tourism Dependence in Latin America. Annals of Tourism Research 10:337–394.
Frank, A. G.
1972 Lumpen-Bourgeoisie and Lumpen-Development: Dependence, Class and Politics in Latin America. New York: Monthly Review Press.
Freeman, D.
1983 Margaret Mead and Samoa: The Making and Unmaking of an Anthropological Myth. Cambridge, MA: Harvard University Press.
Furnham, A.
1984 Tourism and Culture Shock. Annals of Tourism Research 11:41–58.
Furnham, A., and S. Bochner
1986 Social Difficulty in a Foreign Culture: An Empirical Analysis of Culture Shock. *In* Cultures in Contact, S. Bochner, ed., pp. 161–198. Elmsford, NY: Pergamon Press.
Fussell, P.
1982 Introduction to The Road to Oxiana by R. Byron. New York: Oxford University Press.
Gaulis, L., and R. Creux
1975 Pionniers Suisses de l'hotellerie. Paudex, Suisse: Éditions de Fontainemore.
Geertz, C.
1960 The Religion of Java. New York: The Free Press.
Gess, C.
1972 Japanese Travel Habits. Washington DC: U.S. Department of Commerce.
Geva, A., and A. Goldman
1991 Satisfaction Measurement in Guided Tours. Annals of Tourism Research 18:177–185.
Giddens, A.
1979 Central Problems in Social Theory. London: Macmillan.
1983 The Constitution of Society. Oxford: Polity.
1995 Epilogue: Notes on the Future of Anthropology. *In* The Future of Anthropology: Its Relevance to the Contemporary World, A. Ahmed and C. Shore, eds., pp. 272–277. London: The Athlone Press.
Godelier, M.
1980 The Emergence of Marxism in Anthropology in France. *In* Soviet and Western Anthropology, E. Gellner, ed., pp. 3–18. New York: Columbia University Press.
Goldfarb, G.
1989 International Ecotourism: A Strategy for Conservation and Development. Policy Analysis Exercise for John F. Kennedy School of Government, Cambridge, Massachusetts: Harvard University.

Graburn, N.
   1977 Tourism: The Sacred Journey. *In* Hosts and Guests: The Anthropology of Tourism (1st ed.), V. Smith, ed., pp. 17–32. Philadelphia: University of Pennsylvania Press.
   1983a The Anthropology of Tourism. Annals of Tourism Research 10:9–34.
   1983b To Pray, Pay, and Play: The Cultural Structure of Japanese Domestic Tourism. Aix-en-Provence: Centre des Hautes Études Touristiques.
   1989 Tourism: The Sacred Journey. *In* Hosts and Guests: The Anthropology of Tourism (2nd ed.), V. Smith, ed., pp. 21–36. Philadelphia, PA: University of Pennsylvania Press.
   1990 The Hot Spring Inn (*Onsen Ryokan*) as a Symbol of Japanese Neo-Traditionalism. Paper presented at the Conference on Aesthetics and Semiotics: New Visions and New Voices. Los Angeles, CA: University of California.
   1995 The Past in the Present in Japan: Nostalgia and Neo-Traditionalism in Contemporary Japanese Domestic Tourism. *In* Change in Tourism: People, Places, Processes, R. Butler and D. Pearce, eds., pp. 47–70. London: Routledge.
Graburn, N., ed.
   1983 The Anthropology of Tourism. Annals of Tourism Research 10(1).
Graburn, N., and J. Jafari, eds.
   1991 Tourism Social Science. Annals of Tourism Research 16(1).
Gray, H. P., ed.
   1982 The Economics of International Tourism. Annals of Tourism Research 9(1).
Greenwood, D.
   1977 Culture by the Pound: An Anthropological Perspective on Tourism as Cultural Commoditization. *In* Hosts and Guests: The Anthropology of Tourism (1st ed.), V. Smith, ed., pp. 129–138. Philadelphia, PA: University of Pennsylvania Press.
   1989 Culture by the Pound: An Anthropological Perspective on Tourism as Cultural Commoditization. *In* Hosts and Guests: The Anthropology of Tourism (2nd ed.), V. Smith, ed., pp. 171–185. Philadelphia, PA: University of Pennsylvania Press.
Gross, D.
   1984 Time Allocation: A Tool for Studying Cultural Behavior. Annual Review of Anthropology 13:519–558.
Guthrie, G. M.
   1975 A Behavioral Analysis of Culture Learning. *In* Cross-Cultural Perspectives on Learning, R. W. Brislin, S. Bochner and W. Lonner, eds., pp. 96–112. New York: Wiley.
   1981 What you need is Continuity. *In* The Mediating Person: Bridges between Cultures, S. Bochner, ed., pp. 96–115. Boston: Schenkman.
Hallowell, A. I.
   1955 Culture and Experience. Philadelphia, PA: University of Pennsylvania Press.
Hamilton-Smith, E.
   1987 Four Kinds of Tourism? Annals of Tourism Research 14:332–344.
Handler, R.
   1988 Nationalism and the Politics of Culture in Quebec. Madison, WI: University of Wisconsin Press.
   1994 The Authority of Documents at some American History Museums. Journal of American History 18:119–36.

Hardy, D.
  1990 Sociocultural Dimensions of Tourism History. Annals of Tourism
    Research 17:541–555.
Harrell-Bond, B.
  1978 A Window on the Outside World: Tourism and Development in the
    Gambia. American Universities Field Staff Report 19.
  1992 Anthropology and the Study of Refugees. Anthropology Today 8:6–10.
Harris, M.
  1992 Distinguished Lecture: Anthropology and the Theoretical and Paradig-
    matic Significance of the Collapse of Soviet and Eastern Communism.
    American Anthropologist 94:295–305.
Hartmann, R.
  1988 Combining Field Methods in Tourism Research. Annals of Tourism
    Research 15:88–105.
Hassall and Associates
  1992 Comprehensive Tourism Development Plan for Bali, Vols I and II.
    Denpasar, Bali: United National Development Project and Government of
    the Republic of Indonesia.
Haukeland, J.
  1990 Non-travelers: The Flip Side of Motivation. Annals of Tourism Research
    17:172–184.
Heiman, M.
  1989 Production Confronts Consumption: Landscape Perception and Social
    Conflict in the Hudson River Valley. Society and Space 7:165–178.
Herman, S., and E. Schild
  1960 The Stranger Group in a Cross-Cultural Situation. Sociometry 24:165–176.
Hermans, D.
  1981 The Encounter of Agriculture and Tourism: A Catalan Case. Annals of
    Tourism Research 8:462–479.
Herzfeld, M.
  1992 The Social Production of Indifference: Exploring the Symbolic Basis of
    Western Bureaucracy. New York: Berg.
Hofstede, G.
  1980 Culture's Consequences. Beverly Hills, CA: Sage Publications.
Høivik, T., and T. Heiberg
  1980 Centre-periphery Tourism and Self Reliance. International Social Science
    Journal 32:69–98.
Howarth, P.
  1977 When the Riviera was Ours. London: Routledge and Kegan Paul.
Howell, B.
  1994 Weighing the Risks and Rewards of Involvement in Cultural Conservation
    and Heritage Tourism. Human Organization 53:150–159.
Inskeep, E.
  1987 Environmental Planning for Tourism. Annals of Tourism Research
    14(1):118–135.
Ireland, M.
  1990 Come to Cornwall, come to Land's End: A Study of Visitor Experience
    at a Touristic Sight. Problems of Tourism (Problemy Turystyki) 8(3–4):
    33–54.
Iso-Ahola, S.
  1982 Towards a Social Psychological Theory of Tourist Motivation. Annals of
    Tourism Research 9:256–61.

1983 Towards a Social Psychology of Recreational Travel. Leisure Studies 2:45–56.

IUOTO (International Union of Travel Organizations)
1963 The United Nations' Conference on International Travel and Tourism. Geneva: International Union of Travel Organizations.

Jackson, E., and T. Burton
1989 Mapping the Past. *In* Understanding Leisure and Recreation: Mapping the Past, Charting the Future, E. Jackson and T. Burton, eds., pp. 3–28. State College, PA: Venture Publishing.

Jafari, J.
1987 Tourism Models: The Sociocultural Aspects. Tourism Management 8:151–159.
1990 Research and Scholarship: The Basis of Tourism Education. Journal of Tourism Studies 1:33–41.

Jafari, J., and D. Aaser
1988 Tourism as a Subject of Doctoral Dissertations. Annals of Tourism Research 15:407–429.

Jagusiewicz, A.
1990 The Great Mazurian Lakes: Limitations and Prospects of Development. Problems of Tourism (*Problemy Turystyki*) 13:17–34.

Janiskee, R.
1990 Resort Camping in America. Annals of Tourism Research 17:385–407.

JNTO (Japan National Tourist Organization)
1992 Tourism in Japan: 1992. Tokyo: Ministry of Transport.

Johnson, A.
1978 In Search of the Affluent Society. Human Nature 9:51–59.

Jones, P. *et al.*
1983 Understanding Travel Behavior. Brookfield, VT: Gower.

Jordan, J.
1980 The Summer People and the Natives: Some Effects of Tourism in a Vermont Vacation Village. Annals of Tourism Research 7:34–55.

Jurdao Arrones, F.
1990 España en Venta (Segunda Edición). Madrid: Ediciónes Endymion.

Jurdao Arrones, F., and M. Sanchez Elena
1990 España, Asilo de Europa. Barcelona: Editorial Planeta.

Kaplan, J.
1975 The Piaroa. Oxford: The Clarendon Press.

Kardiner, A.
1939 The Individual and His Society. New York: Columbia University Press.
1945 The Psychological Frontiers of Society. New York: Columbia University Press.

Kaur, J.
1985 Himalayan Pilgrimages and the New Tourism. New Delhi: Himalayan Books.

Kelly, J.
1983 Leisure Identities and Interactions. London: Allen and Unwin.

Keyes, C., and P. Van den Berghe, eds.
1984 Tourism and Ethnicity. Annals of Tourism Research 11.

Kitcher, P.
1993 The Advancement of Science. New York: Oxford University Press.

Kottak, C.
1966 The Structure of Equality in a Brazilian Fishing Community. Ph.D. Thesis. New York: Columbia University.

Krippendorf, J.
1986 Tourism in the System of Industrial Society. Annals of Tourism Research 13:517–532.
1987 The Holiday-Makers: Understanding the Impact of Leisure and Travel (V. Andrassy, trans.). London: Heinemann.

Kuper, A.
1994 An Instinct for Compromise. Interview with Aisling Irwin. The Times Higher Education Supplement, 5 August.

Lanfant, M.-F.
1980 Introduction: Tourism in the Process of Internationalization. International Social Science Journal 32(1):14–43.
1993 Methodological and Conceptual Issues Raised by the Study of International Tourism: A Test for Sociology. *In* Tourism Research: Critiques and Challenges, D. Pearce and R. Butler, eds., pp. 70–87. New York: Routledge and Kegan Paul.

Lankford, S.
1992 An Analysis of Resident Preferences, Attitudes and Opinions toward Tourism and Rural Regional Development in the Columbia River Gorge. *In* Joining Hands for Quality Tourism: Interpretation, Preservation and the Travel Industry, R. Tabata, J. Yamashiro and G. Cherem, eds., pp. 225–229. Proceedings of the Heritage Interpretation International Third Global Congress. Honolulu: University of Hawaii, Sea Grant Extension Service.

Laxon, J.
1991 How "We" see "Them": Tourism and Native Americans. Annals of Tourism Research 18:365–391.

Lea, J.
1993 Tourism Development Ethics. Annals of Tourism Research 20:701–715.

Lee, R. B.
1979 The Kung San: Men, Women, and Work in a Foraging Society. Cambridge: Cambridge University Press.

Lee, R. L.
1978) Who Owns Boardwalk: The Structure of Control in the Tourist Industry of Yucatan. *In* Tourism and Economic Change. pp. 19–36. Studies in Third World Societies No. 6. Williamsburg, VA: College of William and Mary, Department of Anthropology.

Leimroth, C., and S. F. Stevens
1984 Peak Experience at the San Fermin. San José, CA: Caislan Press.

Leiper, N.
1992 The Tourist Gaze (review). Annals of Tourism Research 19:604–607.

Lélé, S.
1991 Sustainable Development: A Critical Review. World Development 19:607–621.

Lerner, D.
1958 The Passing of Traditional Society. New York: The Free Press.

Lett, J.
1983 Ludic and Liminoid Aspects of Charter Yacht Tourism in the Caribbean. Annals of Tourism Research 10:35–56.
1989 Epilogue (to T. Nuñez, Touristic Studies in Anthropological Perspective). *In* Hosts and Guests: The Anthropology of Tourism (2nd ed.), V. Smith, ed., pp. 275–279. Philadelphia: University of Pennsylvania Press.

Levy, R.
1983 The Attack on Mead. Science 220:829–832.

1984 Mead, Freeman, and Samoa: The Problem of Seeing Things as They Are. Ethos 12:85–92.

Linton, R., ed.
1963 Acculturation in Seven American Indian Tribes. Gloucester, MA: Peter Smith.

Loukissas, P.
1978 Tourism and Environment in Conflict: the Case of the Greek Island of Myconos. *In* Tourism and Economic Change. pp. 105–132. Studies in Third World Societies No. 6. Williamsburg, VA: College of William and Mary, Department of Anthropology.

MacCannell, D.
1976 The Tourist: A New Theory of the Leisure Class. New York: Schocken.
1989 Introduction (to The Semiotics of Tourism). Annals of Tourism Research 16:1–6.
1992 Empty Meeting Grounds. New York: Routledge and Kegan Paul.

McClosky, M.
1992 Twenty Years of Change in the Environmental Movement: An Insider's View. *In* American Environmentalism: The U.S. Environmental Movement 1970–1990, R. Dunlap and A. Mertig, eds., pp. 77–88. New York: Taylor and Francis.

McCormack, G., and Y. Sugimoto
1988 The Japanese Trajectory: Modernization and Beyond. Cambridge: Cambridge University Press.

McKean, P.
1976 Tourism, Culture Change and Culture Conservation in Bali. *In* Changing Identities in Modern Southeast Asia, D. Banks, ed., pp. 237–247. The Hague: Mouton.
1989 Towards a Theoretical Analysis of Tourism: Economic Dualism and Cultural Involution in Bali. *In* Hosts and Guests: The Anthropology of Tourism (2nd ed.), V. Smith, ed., pp. 119–138. Philadelphia, PA: University of Pennsylvania Press.

Mackie, V.
1988 Division of Labor: Multinational Sex in Asia. *In* The Japanese Trajectory: Modernization and Beyond, G. McCormack and Y. Sugimoto, eds., Cambridge: Cambridge University Press.

Mannell, R., and S. Iso-Ahola
1987 Psychological Nature of Leisure and Tourism Experience. Annals of Tourism Research 14:314–331.

Mansfield, Y.
1992 From Motivation to Actual Travel. Annals of Tourism Research 19:399–419.

Mansperger, M.
1992 Yap: A Case of Benevolent Tourism. Practicing Anthropology 14(2):10–13.
1995 Tourism and Cultural Change in Small-Scale Societies. Human Organization. 54:87–94.

Marris, P.
1975 Loss and Change. Garden City, NY: Anchor Books.

Maslow, A.
1968 Toward a Psychology of Being (2nd ed.). Toronto: Van Nostrand Reinhold.

Mathiesen, A., and G. Wall
1982 Tourism: Economic, Physical and Social Impacts. New York: Longman.

Matthews, H., ed.
1983 Political Science and Tourism. Annals of Tourism Research 10(3).

Matthews, H., and L. Richter
  1991 Political Science and Tourism. Annals of Tourism Research 18: 120–135.
Mazanec, J.
  1995 Constructing Traveler Types: New Methodology for Old Concepts. *In* Change in Tourism: People, Places, Processes, R. Butler and D. Pearce, eds., pp. 137–158. London and New York: Routledge.
Mead, M.
  1949 Coming of Age in Samoa. New York: American Library.
Miller, T.
  1992 Theodore Roosevelt: A Life. New York: Morrow.
Millman, R.
  1988 Just Pleasure: The Churches Look at Tourism's Impacts. Annals of Tourism Research 15:555–558.
Mitchell, L., ed.
  1979 The Geography of Tourism. Annals of Tourism Research 6(3).
Moeran, B.
  1983 The Language of Japanese Tourism. Annals of Tourism Research 10:93–108.
Moffatt, M.
  1989 Coming of Age in New Jersey: College and American Culture. New Brunswick, NJ: Rutgers University Press.
Moon, O.
  1989 From Paddy Field to Ski Slope: The Revitalization of Tradition in Japanese Village Life. Manchester: Manchester University Press.
Moore, K., G. Cushman, and S. Simmons
  1995 Behavioral Conceptualization of Tourism and Leisure. Annals of Tourism Research 22:67–85.
Murdock, G.
  1949 Social Structure. New York: Macmillan.
Murdock, G., C. Ford, A. Hudson, R. Kennedy, L. Simmons, and J. Whiting
  1982 Outline of Cultural Materials (5th ed.). New Haven: Human Relations Area Files.
Nash, D.
  1970 A Community in Limbo. Bloomington, IN: Indiana University Press.
  1976 The Personal Consequences of a Year of Study Abroad. Journal of Higher Education 47:191–204.
  1977 Tourism as a Form of Imperialism. *In* Hosts and Guests: The Anthropology of Tourism (1st ed.), V. Smith, ed., pp. 33–47. Philadelphia, PA: University of Pennsylvania Press.
  1979a Tourism in Pre–Industrial Societies. Aix-en-Provence: Centre des Hautes Études Touristiques.
  1979b The Rise and Fall of an Aristocratic Tourist Culture–Nice: 1763–1936. Annals of Tourism Research 6:61–75.
  1981 Tourism as an Anthropological Subject. Current Anthropology 22: 461–481.
  1984 The Ritualization of Tourism. Annals of Tourism Research 11: 503–506.
  1989 Tourism as a Form of Imperialism. *In* Hosts and Guests: The Anthropology of Tourism (2nd ed.), V. Smith, ed., pp. 37–54. Philadelphia, PA: University of Pennsylvania Press.
  1991 The Course of Sojourner Adaptation: A New Test of the U-Curve Hypothesis. Human Organization 50:283–286.

1992 Epilogue. *In* Tourism Alternatives: Potentials and Problems in the Development of Tourism, V. Smith and W. Eadington, eds., pp. 216–225. Philadelphia, PA: University of Pennsylvania Press.

1993 A Little Anthropology (2nd ed.). Englewood Cliffs, NJ: Prentice Hall.

Nash, D., and R. Butler
1990 Alternative Forms of Tourism. Annals of Tourism Research 18:12–25.

Nash, D., and J. Heiss
1967 Sources of Anxiety in Laboratory Strangers. Sociological Quarterly 8:215–222.

Nash, D., and V. Smith
1991 Anthropology and Tourism. Annals of Tourism Research 18:12–25.

Nash, D., and R. Tarr
1976 The Stranger Group in an Overseas Study Program. The French Review 49:366–373.

Nash, D., and R. Wintrob
1972 The Emergence of Self Consciousness in Ethnography. Current Anthropology 13:527–542.

Nash, R.
1967 Wilderness and the American Mind. New Haven: Yale University Press.

1989 The Rights of Nature: A History of Environmental Ethics. Madison: University of Wisconsin Press.

Neulinger, R.
1974 The Psychology of Leisure. Springfield, IL: Thomas.

1981 To Leisure: An Introduction. Boston: Allyn and Bacon.

Norbeck, E.
1971 Man at Play. Natural History (special supplement):48–53.

Noronha, R.
1979 Paradise Reviewed: Tourism in Bali. *In* Tourism: Passport to Development?, E. de Kadt, ed., pp. 177–204. Oxford: Oxford University Press.

Nuñez, T.
1963 Tourism, Tradition, and Acculturation: Weekendismo in a Mexican village. Southwestern Journal of Anthropology 21:347–352.

1977 Touristic Studies in Anthropological Perspective. *In* Hosts and Guests: The Anthropology of Tourism (1st ed.), V. Smith, ed., pp. 207–216. Philadelphia, PA: University of Pennsylvania Press.

1989 Touristic Studies in Anthropological Perspective. *In* Hosts and Guests: The Anthropology of Tourism (2nd ed.), V. Smith, ed., pp. 265–279. Philadelphia, PA: University of Pennsylvania Press.

Oberg, K.
1960 Culture Shock. New York: Bobbs-Merrill Reprints in the Social Sciences.

Oglethorpe, M.
1984 Tourism in Malta: A Crisis of Dependence. Leisure Studies 3:147–161.

Olwig, K.
1980 National Parks, Tourism, and Local Development: A West Indian Case. Human Organization 39:22–31.

Oppermann, M.
1993 Tourism Space in Developing Countries. Annals of Tourism Research 20:535–556.

O'Rourke, D.
1987 Cannibal Tours. Los Angeles: Direct Cinema Ltd.

Passariello, P.
1983 Never on Sunday? Mexican Tourists at the Beach. Annals of Tourism Research 10:109–122.
Pearce, D.
1986 Tourist Time Budgets. Annals of Tourism Research 15:106–121.
1987 Spatial Patterns of Package Tourism in Europe. Annals of Tourism Research 14:183–201.
1989 Tourist Organizations. New York: Wiley.
1992 Alternative Tourism: Concepts, Classifications, and Questions. *In* Tourism Alternatives, V. Smith and W. R. Eadington, eds., pp. 15–30. Philadelphia, PA: University of Pennsylvania Press.
Pearce, P.
1982a The Social Psychology of Tourist Behavior. New York: Pergamon Press.
1982b Perceived Changes in Holiday Destinations. Annals of Tourism Research 9:145–164.
1990 Farm Tourism in New Zealand: A Social Situation Analysis. Annals of Tourism Research 17:337–352.
1993 Fundamentals of Tourist Motivation. *In* Tourism Research: Critiques and Challenges, D. Pearce and R. Butler, eds., pp. 113–134. London and New York: Routledge and Kegan Paul.
Peil, M.
1977 Unemployment in Banjul: The Farming/Tourist Trade-off. Manpower and Unemployment Research 10.
Pi-Sunyer, O.
1977 Through Native Eyes: Tourists and Tourism in a Catalan Maritime Community. *In* Hosts and Guests: The Anthropology of Tourism, V. Smith, ed., pp. 187–202. Philadelphia, PA: University of Pennsylvania Press.
1989 Changing Perceptions of Tourism in a Catalan Resort Town. *In* Hosts and Guests: The Anthropology of Tourism (2nd ed.), V. Smith, ed., pp. 187–202. Philadelphia: University of Pennsylvania Press.
Picard, M.
1979 Sociétés et tourism: Reflexions pour la recherche et l'action. Paris: UNESCO.
1990 "Cultural Tourism" in Bali: Cultural Performances as Tourist Attractions. Indonesia 49:37–74.
Pigram, J.
1992 Alternative Tourism: Tourism and Sustainable Resource Management. *In* Tourism Alternatives: Potentials and Problems in the Development of Tourism, V. Smith and W. Eadington, eds., pp. 76–87. Philadelphia, PA: University of Pennsylvania Press.
Pizam, A., Y. Neumann, and A. Reichel
1978 Dimensions of Tourist Satisfaction with a Destination Area. Annals of Tourism Research 5:314–322.
Preister, K.
1987 Issue-centered Social Impact Assessment. *In* Anthropological Praxis: Translating Knowledge into Action, R. Wulff and S. Fiske, eds., pp. 39–55. Boulder: Westview Press.
Pye, E., and T.-B. Lin
1983 Tourism in Asia: The Economic Impact. Singapore: Singapore University Press.
Quiroga, I.
1990 Characteristics of Package Tours in Europe. Annals of Tourism Research 17:185–207.

Rajotte, F., and R. Crocombe, eds.
   1980 Pacific Tourism: As Islanders See It. South Pacific Social Sciences Association and The Institute of Pacific Studies. Fiji: University of the South Pacific.
Rapport, N.
   1993 Diverse World Views in an English Village. Edinburgh: University of Edinburgh Press.
RDCNTA (Research Department, National Tourism Administration, People's Republic of China)
   1989 Tourism and Change in Life Style: The China Case. Problems of Tourism (*Problemy Turystyki*) 12:65–70.
Riemer, G.
   1990 Packaging Dreams. Annals of Tourism Research 17:501–512.
Ritchie, J. R. B.
   1992 Tourism Research: Policy and Managerial Priorities for the 1990's and Beyond. *In* Tourism Research: Critiques and Challenges, D. Pearce and R. Butler, eds., pp. 201–216. New York: Routledge and Kegan Paul.
Rojek, C.
   1985 Capitalism and Leisure Theory. London and New York: Tavistock.
   1989 Leisure and Recreation Theory. *In* Understanding Leisure and Recreation: Mapping the Past. Charting the Future, E. Jackson and T. Burton, eds., pp. 69–88. State College, PA: Venture Publishing.
   1993 Disneyculture. Leisure Studies 12:121–135.
Rosaldo, R.
   1984 Toward an Anthropology of Self and Feeling. *In* Culture Theory, R. Shweder and R. Levine, eds., pp. 137–157. Cambridge: Cambridge University Press.
Rosenberg, H.
   1988 A Negotiated World. Toronto: University of Toronto Press.
Sage, C.
   1989 Extending Service: Sheraton Woos the Japanese Traveler. *In* World Class Service, G. Shames and G. Glover, eds., pp. 145–154. Yarmouth, ME: Intercultural Press.
Saglio, C.
   1979 Tourism for Discovery: A Project in Lower Casamance. *In* Tourism: Passport for Development?, E. de Kadt, ed., pp. 321–338. Oxford: Oxford University Press.
Sahlins, M.
   1972 Stone Age Economics. Chicago: Aldine.
Sale, K.
   1993 The Green Revolution: The American Environmental Movement, 1962–1992. New York: Hill and Wang.
Sayers, S.
   1989 Work, Leisure, and Human Needs. *In* The Philosophy of Leisure, T. Winnifrith and C. Barrett, eds., pp. 34–53. New York: St Martin's Press.
Schlechten, M.
   1978 Aliénation ou rencontre: modèles de tourisme culturel en Afrique. Thèse, License. Fribourg, Suisse: Université de Fribourg Suisse.
   1988 Tourisme Balnéaire ou Tourisme Rural Intégré? Deux Modèles de Développement Sénégalais. Saint-Paul Fribourg Suisse: Éditions Universitaires Fribourg Suisse.

Schooler, C.
1972 Social Antecedents of Adult Psychological Functioning. American Journal of Sociology 78:299–322.
Schuetz, A.
1944 The Stranger: An Essay in Social Psychology. American Journal of Sociology 49:499–507.
1945 The Homecomer. American Journal of Sociology 50:369–376.
Selwyn, T.
1990 Tourist Brochures as Post-Modern Myths. Problems of Tourism (*Problemy Turystyki*) 8:13–26.
Seton, E.
1910 Boy Scouts of America: A Handbook of Woodcraft, Scouting, and Lifecraft. New York: Boy Scouts of America.
Shames, G., and G. Glover, eds.
1993 World Class Service. Chicago: Intercultural Press.
Shannon, T. R.
1989 An Introduction to the World System Perspective. Boulder: Westview Press.
Silberbauer, G.
1972 The G/Wi Bushmen. *In* Hunters and Gatherers Today, M. Bicchieri, ed., pp. 271–326. New York: Holt, Rinehart and Winston.
Silver, L.
1993 Marketing authenticity in Third World Countries. Annals of Tourism Research 20:302–318.
Simmel, G.
1950 The Sociology of Georg Simmel, K. Wolff, ed., Glencoe, IL: The Free Press.
Singh, T. V.
1989 The Kulu Valley: Impact of Tourism Development in the Mountain Areas. New Delhi: Himalayan Books.
Smith, T.
1973 The Policy Implementation Process. Policy Sciences 4:197–209.
Smith, V.
1979 Women: The Taste Makers in Tourism. Annals of Tourism Research 6:49–60.
1981 Controlled vs. Uncontrolled Tourism: Bhutan and Sikkim. Rain 40:4–6.
1989 Eskimo Tourism: Micro Models and Marginal Men. *In* Hosts and Guests: The Anthropology of Tourism, V. Smith, ed., pp. 55–82. Philadelphia, PA: University of Pennsylvania Press.
1992 Managing Tourism in the 1990's and Beyond. Practicing Anthropology 14:3–4.
Smith, V., ed.
1977 Hosts and Guests: The Anthropology of Tourism (1st ed.). Philadelphia, PA: University of Pennsylvania Press.
1989 Hosts and Guests: The Anthropology of Tourism (2nd ed.). Philadelphia, PA: University of Pennsylvania Press.
Smith, V., and W. Eadington, eds.
1992 Tourism Alternatives: Potentials and Problems in the Development of Tourism. Philadelphia, PA: University of Pennsylvania Press.
Smith, V., A. Heatherington, and M. Brumbaugh
1986 California's Highway 89: A Regional Tourism Model. Annals of Tourism Research 13:415–434.

Spill, J.-M.
1976 Les charters en Méditerranée. Annales de Géographie 468:206–224.
Spindler, G., and L. Spindler
1990 The American Cultural Dialogue and its Transmission. London: The Falmer Press.
Spiro, M.
1992 Anthropological Other or Burmese Brother? New Brunswick, NJ: Transaction Publishers.
Srisang, K.
1989 The Ecumenical Coalition on Third World Tourism. Annals of Tourism Research 16:119–121.
SSRC Seminar (Social Science Research Council)
1954 Acculturation: An Exploratory Formulation. American Anthropologist 56:973–1002.
Stabler, M.
1992 Where will the Birds Go? The Threats of Recreational and Tourism Development to Estuarial Wildlife: An Economic Critique. *In* Proceedings of the Heritage Interpretation International Third Global Congress, R. Tabata, J. Yamashiro, and G. Cherem, eds., pp. 367–370. Honolulu: University of Hawaii, Sea Grant Program.
Stevens, S.
1993 Claiming the High Ground: Sherpas, Subsistence, and Environmental Change in the Highest Himalayas. Berkeley: University of California Press.
Strathern, M.
1981 Kinship at the Core. Cambridge: Cambridge University Press.
Stringer, P., ed.
1984 The Social Psychology of Tourism. Annals of Tourism Research 11(1).
Suomi, S.
1982 Why does Play Matter? The Behavioral and Brain Sciences 5:169–170.
Sutton, W.
1967 Travel and Understanding: Notes on the Social Structure of Touring. International Journal of Comparative Sociology 8:218–223.
Swianiewicz, M., and P. Swianiewicz
1989 The Catholic Church in Poland as the Organizer of Youth Tourism. A New Organizer of Old Forms or an Alternative Model. Problems of Tourism (*Problemy Turystyki*) 3:50–58
Thomas, W., and F. Znaniecki
1927 The Polish Peasant in Europe and America. New York: Knopf.
Towle E., and J. Towle
1991 Environmental Agenda for the 1990's. Bridgetown, Barbados: Caribbean Conservation Association, Island Resource Foundation.
Towner, J.
1985 The Grand Tour: A Key Phase in the History of Tourism. Annals of Tourism Research 12:297–334.
1988 Approaches to Tourism History. Annals of Tourism Research 15:47–62.
Tsartas, P.
1992 Socioeconomic Impacts of Tourism on Two Greek Isles. Annals of Tourism Research 19:516–533.
Turnbull, C.
1962 The Forest People. New York: Touchstone Books.
Turner, L., and J. Ash
1975 The Golden Hordes. London: Constable.

Turner, V.
 1969 The Ritual Process. Chicago, IL: Aldine.
Turner, V., and E. Turner
 1978 Image and Pilgrimage in Christian Culture. New York: Columbia University Press.
United Nations
 1963 Recommendations on International Travel and Tourism. Geneva: United Nations.
Urry, J.
 1990 The Tourist Gaze. London: Sage Publications.
van der Wurff, A., M. Wasink, and P. Stringer
 1988 "In situ" Recording of Time-Sampled Observations. Annals of Tourism Research 15:166–172.
van Gennep, A.
 1960[1908] The Rites of Passage. Chicago: University of Chicago Press.
Wagner, U.
 1977 Out of Time and Out of Place-Mass Tourism and Charter Trips. Ethnos 42:38–52.
Wahab, S.
 1975 Tourism and Air Transport. Tourist Review 28:146–151 and 29:9–11.
Wall, G.
 1992 Tourism Alternatives in an Era of Global Climatic Change. *In* Tourism Alternatives: Potentials and Problems in the Development of Tourism, V. Smith and W. Eadington, eds., pp. 194–215. Philadelphia, PA: University of Pennsylvania Press.
Wallace, A.
 1978 Rockdale. New York: Knopf.
Wallerstein, I.
 1974 The Modern World System I. New York: Academic Press.
 1994 Comment. Current Anthropology 35(1):9–10.
Watson, G. L., and J. Kopachevsky
 1994 Interpretations of Tourism as Commodity. Annals of Tourism Research 21:643–660.
Weigle, M.
 1989 From Desert to Disneyworld: The Santa Fe Railway and the Fred Harvey Company Display the Indian Southwest. Journal of Anthropological Research 45:115–137.
Wolf, E.
 1982 Europe and the People Without History. Berkeley: University of California Press.
WCED (World Commission on Environment and Development)
 1987 Our Common Future. New York: Oxford University Press.
WTO (World Tourism Organization)
 1986 Bhutan Final Report. Tourism Development Market Plan (BHU/81106). Madrid: World Tourism Organization.
 1993 Sustainable Tourism Development: a Guide for Local Planners. Madrid: World Tourism Organization.
Worsley, P.
 1984 The Three Worlds. Chicago: University of Chicago Press.
Yiannakis, A., and H. Gibson
 1992 Roles Tourists Play. Annals of Tourism Research 19:287–303.

Young, R.
  1977 The Structural Context of the Caribbean Tourist Industry: A Comparative Study. Journal of Rural Cooperation 5:657–672.
Zonabend, F.
  1980 La mémoire longue: Temps et histoires au village. Paris: Presses Universitaires de France.

Author Index

Subject Index